Taking the Wheel Out: A Deep Dive into Autonomous Vehicles

Johann

Contents

Chapter 1

General Introduction

Automated driving technology is one of the key areas of research and development in the automotive industry to date. The technology promises to bring vast improvements in terms of safety, convenience, and accessibility of transport. In the long run, it is expected to help optimize traffic flow, increase energy efficiency, and, in making car sharing highly convenient, dramatically reduce the number of cars and parking space needed in cities, as well as drastically reducing the overall cost of transportation. Automated driving technology is thus set to completely change the way we use cars, transform our cities, and to have a great impact on our day-to-day lives.

Before we get to this point, however, there are a number of challenges ahead of us on multiple fronts. In order to guarantee the highest standard of safety for automated vehicles (AVs) in most or all conditions, major improvements need yet to be made to the technology itself. Furthermore, a wide array of legislative hurdles have to be crossed, many of them tightly interwoven with open ethical questions.

This book is concerned with these issues both on the ethical and technological side, containing four publications concerned with ethical decision making in dilemmatic situations in road traffic, as well as one publication on improvements for deep neural networks – the core technology behind the rapid progress in object and scene recognition in recent years and a driving force in the development of AV technology.

1.1   SAE levels of automation

Before outlining the technology in development for automated vehicles, we will briefly visit the functional topology, or levels, of automated driving. The Society of Automotive Engineers's (SAE) has introduced a categorization of automated driving functions in six levels from no automation (level 0) to fully automated (level 5) [102].

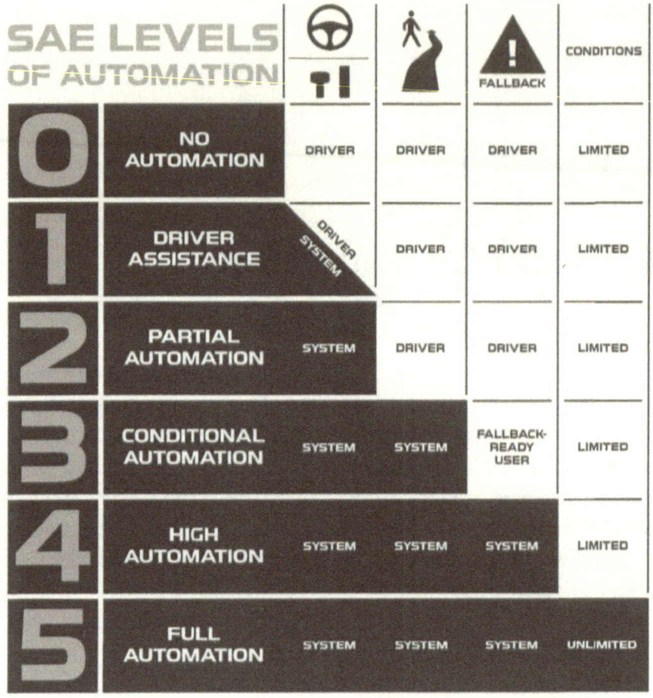

Figure 1.1: Overview: SAE levels of automation. Columns left to right: Sustained lateral and longitudinal vehicle motion control, object and event detection and response (OEDR), dynamic driving task fallback, operational design domain (ODD).

Determining factors for the categorization of a car's Advanced Driver Assistance Systems (ADAS) or Automated Driving (AD) features are the partial or complete handling of the dynamic driving task (DDT) and DDT-fallback, as well as the operational driving domain (ODD), i.e., the conditions like traffic, weather, time-of-day, etc., under which these tasks can be performed. The dynamic driving task includes *sustained lateral and longitudinal vehicle motion control* and *object and event detection and response* (OEDR), and fallback refers to the task of achieving a so-called *minimal risk condition*, in which the risk of a collision is minimized.

On level 0, all of the relevant tasks are performed by the driver. The car can be equipped with active safety features, but will still be categorized as level 0 if these don't control the lateral (steering) or longitudinal (acceleration and braking) motion control on a sustained level. Level 1 describes systems that either take over the steering, or acceleration and braking for sustained periods of time, while level 2 describes systems that perform both of these tasks, but still leave the OEDR to the driver.

Level 3 systems add OEDR to the mix, but may rely on a fallback-ready user, who, upon request, is tasked with achieving the minimal risk condition (DDT fallback). On level 4, the DDT fallback is fully handled by the system, and the only limit to the automated driving capabilities is the restriction to certain conditions. In a full-fledged automated driving system, i.e., level 5, these restrictions are lifted, and the system can operate the car in any condition in which a human could do the same, i.e., regardless of the environment or type of road, weather or time-of-day conditions, and so forth.

While the first two levels of automation are widely available on the market in 2020, level 3 systems are still very rare. Systems up to level 2 may rely on constant supervision by a human driver, while level 3 requires OEDR to be performed by the system. For many environments and conditions, a level 3 system is thus required to host a fully developed object detection stage, and needs to be programmed to appropriately handle unexpected events. These requirements are arguably the most difficult to achieve from a technical standpoint. Many systems at level 3 and higher will also require specific regulation, as the driver is no longer tasked with OEDR, and thus no longer fully responsible for the actions of their vehicle.

In 2019, Audi released the first commercially available level 3 system in form of their Traffic Jam Pilot, allowing the driver to engage in non-driving activities during traffic jams [3]. Notably, this system works exclusively on highways and up to a speed of 60 kph, at which point it will hand over control to the driver. Limiting the systems to traffic jams on highways considerably relaxes the requirements for object detection and event response, as it constitutes a highly predictable driving environment. The Traffic Jam Pilot thus avoids the most demanding operational domains, such as cities, rural roads, and high velocity OEDR on highways. Since these driving environments constitute a significantly tougher challenge with respect to technology, ethics, and legislation, it is yet unclear when automated driving in these domains will be achievable.[1]

---

[1]Notably, as of the time of writing this book, Tesla has begun rolling out beta-versions of their "Full Self-Driving" (FSD) software to select owners of Tesla vehicles equipped with the required hardware. While this system represents likely the most advanced driving aid in commercially available vehicles to date, it is, despite its name, a level 2 system that requires constant supervision by the driver on board.

In this book, we focus on the latter, not yet achieved driving domains and challenges that go alongside them.

1.2   Control Methodology in Automated Vehicles

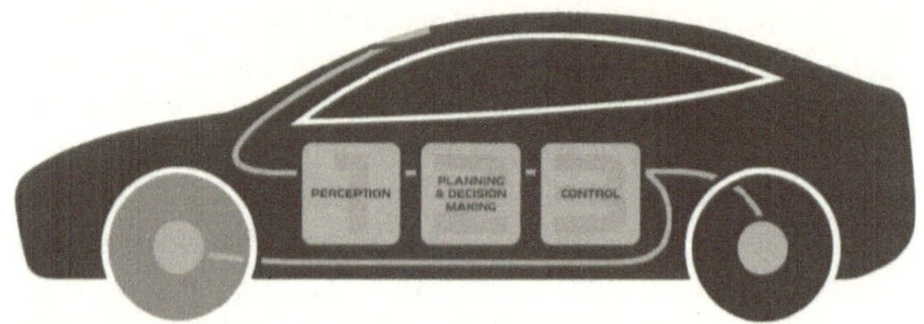

Figure 1.2: Simplified illustration of the signal flow in sense-plan-act control architectures widely used in (partially) automated vehicles.

In order to provide some more context for the publications in this book, consider the sense-plan-act model, the standard control methodology for advanced driver-assistance systems (ADAS) and automated driving systems (ADS) [103]. This architecture defines the flow of information processing in AVs from sensors to actuators, and consists of three main stages, as detailed in Gruyer et al. [60]:

1. The *perception stage* collects, filters and processes data from all sensors and systems, as well as information from external sources such as other vehicles, road infrastructure, and maps. Functionally, the tasks of the perception system are to detect the road, lanes, relevant features of the road and obstacles, detect or sense relevant information about the environment, such as weather, road signs, etc., and to correctly apprehend the vehicle's position and dynamic state. This also includes behavior identification and trajectory prediction for moving objects in the scene. In a process referred to as sensor fusion, the information derived from various sources is then combined in a local map of the car's immediate surrounding, in Gruyer et al. [60] referred to as the local dynamic perception map (LDPM).

2. The *planning and decision-making stage* uses the LDPM and route information to select the car's path, generate precise trajectories for the AV and make tactical decisions, such as the desired speed,

acceleration or deceleration rates. Ad-hoc risk analysis, risk management and ethical assessments, and the subsequent decision making happen within this module.

3. Finally, the *control stage* issues precise control orders to the steering system, engine or motors, and brakes that ensure that the path and other tactical decisions, decided on in the planning stage, are effectuated by the vehicle.

This three-tiered structure follows a logical perception-to-action sequence, and allows for the individual sub-tasks, i.e., perception, planning, and action, to be verified and regulated separately. In doing so, it also provides a leverage point for regulation.

1.3 Sensors, Perception, and Deep Neural Networks in Automated Vehicles

To perceive their environment for automated driving functions, AVs are typically equipped with a large number of sensors, most commonly cameras, radar, and LiDAR (light detection and ranging) systems [64]. Radar provides precise data about the relative location and speed of other objects, and is largely indifferent to environmental conditions. However, spatial resolution is low, and radar systems are not able to perform object recognition. For this purpose, they are typically combined with camera systems providing classification labels for the detected objects, such as other cars, pedestrians, and road signs. Both camera and radar are relatively inexpensive systems, yet in combination still suffer from the shortcomings of cameras in low light and adverse conditions. LiDAR systems are based on laser sensors, and do not rely on ambient light, providing high resolution 3D maps of the vehicles's environment. While the resolution is not as high as that of camera systems, and they lack color and brightness information, they could, in the long term, replace radar technology and could be used in conjunction with cameras. For the time being, however, the cost of LiDAR systems is too high to be economically viable for regular cars, and while widespread in AV prototypes, the first generation of automated vehicles is expected to mostly rely on a combination of radar and camera systems.

Due to the reliance on camera systems for important tasks such as object and lane recognition in the foreseeable future, the respective algorithms are of vital importance for AVs. The rapid progress that has been achieved on this front in recent years has largely been driven by advances in deep neural networks, and in particular convolutional neural networks (CNNs) for image processing. In simplified terms, CNNs perform feature extraction from images by alternating between the application of linear filters (convolution) and non-linear functions, called activation functions. The use of non-linear functions is necessary for the extraction of complex, non-linear features, and their choice can have a large impact on the networks' performance. As part of this book, we investigate the option of replacing the commonly static and pre-defined activation functions in CNNs with parametric functions that are learnt by the system during training, called Adaptive Blending Units (ABUs). In the context of automated vehicles, ABUs could help to improve detection and correct classification rates.

1.4   Ethical Questions Concerning Planning and Decision Making

The majority of this book is concerned with the vehicle's ad-hoc decision making in response to unexpected events and suddenly occurring obstacles. The central question here is how the system should decide between several undesirable outcomes with no good option available. A common example is that of a pedestrian crossing the street unexpectedly right in front of the car, where the only way to avoid a collision with the pedestrian is to swerve and collide with an oncoming car. Variations of this paradigm include the juxtaposition of pedestrians of different age and gender, animals, and the option to sacrifice the car's occupants in favor of a pedestrian or group of pedestrians. These types of situation are commonly referred to as dilemma situations, due to their apparent similarity to dilemmatic thought experiments, such as the trolley dilemma[117]. [2]

To obtain a point of reference for what constitutes "right" behavior when an AV is facing a dilemma situation, we chose to investigate how we as people behave in these situations. In the first publication of this book, we used a forced choice paradigm in an immersive virtual reality

---

[2]In the trolley dilemma, a runaway trolley is headed for a group of five people, but can be diverted onto another track where it would kill only a single person. The central dilemma here arises from the clash of two moral maxims; doing what's best for the greater good (saving the greater number of people) and not actively causing harm or killing a person. It is debatable to what extent a clash of moral maxims is also present in many of the situations that are discussed in the realm of AVs, or whether these merely ask for a valuation of personal features and other factors relevant to the decision making. Nonetheless, we use the term *dilemma situations* to mean any situation in which the lesser of to evils has to be determined.

(VR) setting to analyze the central tendencies in participant's behavior, and explore options of modelling this behavior. The study was well received by the media and sparked some debate in the scientific community, leading to a commentary article by Keeling [75]. The second publication in this book is, therefore, a response to Keeling's commentary, in which we discuss questions regarding the methodology, and give our take on the relevance of the study with respect to the car's programming, legislation, and public communication of the former. Completing this section of the book, in the third publication, we again used a forced choice paradigm. This time, we analyzed the impact that different assessment methodologies may have on the participant's behavioral choices, ranging from abstract questionnaire-style assessments, to immersive VR settings. We believe that these studies are a valuable contribution to this field, shedding light on different aspects – methodology of assessment, modeling of behavior, and assessment of central behavioral tendencies. Together with a number of publications by colleagues from within and outside of our lab (e.g., Awad et al. [5], Bergmann et al. [11], Faulhaber et al. [39]), these studies contribute to a rather clear picture of what behavior we as humans intuitively deem morally right, or acceptable, in such dilemma situations.

While clear-cut dilemma situations are arguably very rare occurrences in the real world, the issue of having to prioritize between a number of undesirable potential outcomes, or between conflicting behavioral maxims is going to be faced by any AV on a day-to-day basis. AVs will have to perform some form of risk management in order to reasonably adapt their driving speed to their environment and the driving situation, or to determine when it is safe enough to overtake a cyclist or another car. In other situations, traffic regulations need to be violated, for example when the only way to let an ambulance pass is to run a red light. Doing so, however, requires the car to make sure that the risk to other road users is kept within acceptable bounds. But issues like these are not simply a question of the right programming, they need to go hand in hand with legislation. We analyzed these issues and summarized the current state of the debate in the fourth publication of this book. In this, we derive a list of ten demands and open questions that need to be addressed in order to create a legislative framework for ethical decision making in AVs.

1.5 Outline

In the following chapter, we will take a deeper dive into the related literature, extract open questions and a working hypothesis, before presenting the publications in the previously stated order. All publications are prefaced with a short laymen's summary, putting them into context and outlining key aspects and achievements.

Chapter 2

Ethical Decisions in Road Traffic: A Deeper Dive

In this chapter, we will define and explain key terms surrounding the topic of automated ethical decision making in road traffic, portray the relevant literature that delivered the basis for our research, lay out the motivation behind this book and derive the research objective from this.

2.1 Dilemmas in Literature and Practice

First, we will discuss various types of dilemmas in literature and practice, i.e., classic dilemmas and thought experiments, traffic dilemmas, and a number of further dilemmas occurring in the real world.

2.1.1 The Trolley Problem and other Classic Dilemmas

Much of the research in behavioral ethics has been based on thought experiments and hypothetical dilemma situations, such as the Trolley Dilemma, which was first proposed by [117]. In this, a runaway trolley is headed toward a group of five people. The only way to save these five is to pull a lever and divert the trolley onto a different track, killing a single person instead. The agent is thus faced with the decision of letting five people die, or sacrificing a single person instead. While both options are unfavorable, the core of this dilemma is the notion that by pulling the lever, the agent becomes an active part of the equation; the situation thus emphasizes a distinction between an active killing and a passive letting die. Consequentialist and utilitarian moral approaches aim to minimize the overall harm (or maximize overall utility), and would favor the single individual to be sacrificed in favor of the group of five. Deontological approaches instead focus on the action, rather than the consequences, and would typically deem the active interference that leads to the killing of a person as unacceptable, thus favoring inaction. The level of personal and active involvement is further increased in a variation of the Trolley Dilemma, called the Footbridge Dilemma [118]. In this, instead of pulling a lever, the agent has to physically push a large man off a bridge and in front of the trolley to stop it from killing the group of five. Beyond these, there are a variety of structurally similar dilemmas used in the literature, such as the fumes dilemma, the crying baby dilemma, and many more (see [57] supplement for a large collection of high and low stakes dilemmas). What all of these dilemmas typically have in common is that they construct situations in which fundamental ethical schools of thought clash and propose opposing actions. While these thought experiments are often quite remote from the real world with no direct relevance to commonly occurring events, they serve well to research and discuss our fundamental moral intuitions.

2.1.2 Traffic Dilemmas

When we talk about dilemma situations in the context of road traffic, however, we do not focus on the clash of moral schools of thought. Instead, we refer to any situation in which an ad-hoc decision has to be made that involves some level of risk to the health and well-being of humans and animals as a dilemma situation. This naturally entails situations in which a collision is unavoidable, but a choice exists with regards to who is protected and who is sacrificed. Beyond this, our definition also includes situations with very small levels of risk, such as deciding if and when to overtake a cyclist (i.e., when is the risk of overtaking low enough to be acceptable, given the reduction in travel time for the vehicle). Moreover, we include situations in which traffic regulations are in conflict with the prevention or mitigation of harm to people, property damage, or any risk thereof. This includes situations in which, for example, a red light needs to be jumped in order to let an ambulance pass, or a solid lane marking has to be crossed to avoid a collision. We deem the inclusion of small, and possibly unknown levels of risk, as well as traffic regulations an important part of this definition, as situations involving these factors occur frequently in traffic, and can therefore not be regarded "too rare to matter". The notion of a dilemma in these situations arises from the need to weigh utility and conformity with traffic regulations against (often) low levels of risk for bodily harm.

Note that an ethical dilemma situation doesn't lose its ethical relevance because a simplistic system is used to implement a moral framework. For instance, if a system is programmed to follow the hard rule of never swerving into the opposite lane or onto the sidewalk, then it merely implements a moral framework that favors the literal interpretation of traffic regulations above all else. This has no bearing on the ethical relevance of the situation, so long as the actor (i.e., the driver or ADS) has the physical ability to change the outcome, and a perception or comprehension mechanism connecting actions with foreseeable consequences.

The concepts of utilitarian and deontologically based ethics also appear in the context of road traffic, for example in the question if and when one person can be harmed or put at risk at the benefit of another person, or a group of others. However, many of the open questions in this realm only remotely involve a juxtaposition of different schools of thought, and rather revolve around valuations of utility, harm, rule violations, etc. These questions are thus reflected in this book' behavioral studies (publications 1 and 3), and brought up again in the comprehensive discussion of this topic in publication 4.

2.1.3 Parallels to Dilemmas in Other Fields

Beyond classical thought experiments and traffic dilemmas, we find practically relevant dilemmas, for example, in the conservation of biodiversity [126], or, in the field of medicine, in organ recipient rankings [78] and the triage process [34]. These dilemmas typically result from a severe lack of resources such as funding, time, available personnel, medical equipment, drugs, or doner organs. Similar to traffic dilemmas, such cases can require a priority ranking of the individuals at risk. Prior research and philosophical debate exists on these issues, and can be borrowed from to inform discussions and research in traffic dilemmas.

For example, in debates about the conservation of biodiversity, [126] identified three moral principles, proponents of which appear to be in a stalemate; The principle of equality, assigning equal value to each species, but disregarding their functional roll as parts of ecosystems. The principle of utility, aiming to achieve the greatest good for the greatest number of species, but still allowing for some species to be sacrificed, and suffering from an unclear definition of "good". And the principle of need, giving priority to the species in most immediate danger. Arguably, versions of these three principles also apply to dilemmas in road traffic. For example, the principle of need could be understood as prioritizing the party that is at risk for the most severe bodily harm, regardless of other factors, while the principle of utility could factor in a variety of values, such as protection of pedestrians, adherence to traffic regulations, etc. As a way of approaching the opposition to triage processes on the basis of non-utilitarian ethical principles, the authors argue that, amongst others, it is necessary to communicate the triage objectives and protocols openly, consider the risk preferences of various stakeholders, and to regularly review and adapt the protocols in place. Adaptation of these principles for traffic dilemmas is straight forward, and they are mirrored in the demands stated in publication four of this book, where we aim to develop a basis for a framework of ethical decision making in traffic dilemmas.

In the medical field, [15] have analyzed the impact of patient features such as age, sex, or parenthood on how participants ranked them on hypothetical organ recipient lists. Many of the same questions that are being discussed in the context of traffic dilemmas, such as whether or not younger people should receive preferential treatment. With regards to the use of their findings, the authors state that they do not suggest to take empirically assessed community values as a model for organ recipient rankings, but to use these to inform the debate – a view we can adapt for observed behavioral patterns of participants in traffic dilemmas.

Beyond this, the medical field also offers a model for priority rankings. In the US, organ recipient rankings are created using a point-system allocating points "for the degree of antigen matching, the waiting time on the transplant list, immunologic sensitisation and medical urgency" [15], and similar systems are in place in many other countries as well. These systems reflect a weighing of medically relevant criteria, and exclude discrimination based on factors relating to the identity, such as age, sex, ethnicity, or other group affiliations. This principle of weighing a number of relevant factors (and precluding others from consideration) is established practice in the field of medicine around the globe and has been applied for decades without significant backlash. Thus, a similar system could be taken into consideration for the traffic sector as well.

2.2 Theories of Moral Judgment and Behavior

In order to describe moral judgment and behavior, researchers typically aim to construct models that explain or predict the responses of participants facing moral dilemmas. These models take into account key descriptive features of the situation in question, as well as, in some cases, features of the participants. Such models can be roughly divided into two categories; qualitative or explanatory models, postulating certain mental processes, and quantitative or statistical models, typically without assumptions about hidden processes. In this section, we will focus on models and theories aiming to explain mental processes, as it represents the dominant approach in behavioral ethics outside of AV-related studies.

Most prominently, the dual process theory, put forward by [57], proposes two distinct cognitive systems in competition. The first system is a fast, intuitive and emotionally rooted system. It is thought to elicit negative affect when behavioral rules are violated, essentially making us feel bad in anticipation of certain behavior, and thus preventing us from engaging in it. As it is focusing on the moral valence of behavior and not on consequences, this system favors decisions in line with deontological ethics. In contrast, the second system is a slower, controlled and reasoning-based system. It performs cognitive deliberation of the consequences of a decision and thus favors decisions in line with utilitarian ethics. Within the dual process theory, varying responses between the trolley dilemma and the footbridge dilemma are explained by the more personal and, therefore, emotionally engaging action of pushing a man off a bridge versus merely pulling a lever, as well as the use of harm as means to an end. These aspects are said to shift focus towards the intuitive system, resulting in higher endorsement of inaction over action, compared to the original trolley dilemma [56].

However, the dual-process theory, based on the emotion-cognition distinction is not undisputed. [28] argues that while a distinction between competing processes is necessary, the distinction between affective and non-affective processing is inadequate, since both processes must involve both cognition and affective content. Instead, he proposes a distinction based on two cognitive mechanisms borrowed from the field of reinforcement learning. The first is an action-based system, assigning reward values to possible actions in a given situation. These reward values are learned from experience and statically assigned to a given situation-action-pair. The second mechanism is outcome-based and relies on an underlying world model. In simplified terms, it predicts the consequences of the possible actions in a given situation and reassigns the value of the consequence to the action that leads to it. In the trolley dilemma, the outcome-based system would favor utilitarian behavior, and the action-based system would not intervene because the action of pulling a lever is not generally associated with negative reward. Conversely, the action of pushing a person off a bridge is associated with negative reward, thus explaining the lower endorsement rates of utilitarian behavior in the footbridge dilemma. Further evidence in favor of the action vs. outcome distinction in dual-process models is given, e.g., in [30] and [41].

In another theory in the realm of moral judgment, [62] aim to explain different views of opposing political camps (liberals and conservatives) with a model of morality based on five factors, and the relative importance of each of these factors to members of the political camps. Finally, based on a large body of neuroscientific evidence, [93] propose a detailed account of moral emotions as the foundation of our moral judgment. While none of the two entail concrete predictions with respect to moral decision-making in the trolley dilemma and similar scenarios, they demonstrate that the scope of the dual-process theories is limited, and that we are a long way from a comprehensive theory about the cognitive mechanisms governing our moral judgment and behavior.

2.3 Factors Influencing Moral Decisions

Several factors influencing participant behavior in dilemma studies have been identified prior to the bulk of studies ranging around moral dilemmas in traffic scenarios. The framing of a dilemma as more personal (e.g., "I would do..." instead of "it is acceptable to..."), or in a negative manner (e.g., focusing on number of victims instead of survivors) was found to decrease utilitarian responses [56, 19]. Increased emotional proximity to potential victims was similarly found to decrease utilitarian responses [115]. Touching on both the aspects of framing and emotional proximity, dilemmas presented in a language other than the participants' first language was shown to lead to stronger utilitarian endorsement, possibly moderated by a greater psychological distance [26]. Neuroscientific evidence further supporting the link between emotional engagement and response patterns is provided by [114], who showed that the utilitarian responses become more prevalent when the right dorsolateral prefrontal cortex, associated with emotional processing, is disrupted. A person's affective state was also found to influence response patterns. In [121], experimentally induced positive affect lead to more utilitarian responses. The authors suggest that negative affect connected to rule violations may be cancelled out by the experimentally induced positive affect. Another factor with an impact on moral decision making is response time. [58] found that under increased cognitive load, response times for utilitarian responses were longer than for non-utilitarian responses. This suggests that additional time for processing may be needed to override intuitive non-utilitarian response tendencies. Similarly, in [97], persuasive arguments were shown to have an impact on moral judgments, but only when additional time for deliberation was given. Finally, [108] varied response time windows systematically and, in line with previous observations, found longer deliberation to shift response patterns towards more utilitarian responses. More recently, the community has begun looking at the psychological traits as a basis for utilitarian choices more closely [25, 38]. Studies found that utilitarian choices are preferred by people with reduced aversion to harm [30], those aiming to maximize their own self-interest [73], and those with higher scores in psychopathy [79, 9, 73]. [98] explain the effects of trait psychopathy on moral actions with a reduced emotional reactivity to harmful acts, tying in the above findings with those on emotional proximity. Moreover, how often people have to make dilemmatic decisions in real life may influence their r esponses. For example, [24] found military pilots, who make more morally ambiguous decisions than other military personnel, to make fewer utilitarian decisions in the text-based dilemmas presented to them, than other military personnel. People in helping professions were found to make fewer utilitarian decisions than others in some studies [59, 101], but showed the same response patterns as a control group in [42]. These findings d eliver a b ackdrop for the broader analysis of influencing f actors i n r oad t raffic sc enarios in publications one and three.

2.4 Methods of Assessment in Behavioral Ethics

Encompassed in the term *assessment methodology* are both the presentation of a dilemma, as well as the way participants enter their response. The most prevalent modes of presentation in the literature on behavioral ethics are text-based descriptions of dilemma situations (e.g., [56, 83, 71, 13]), illustrations such as the birds-eye view drawings of the scenarios in question in the Moral Machine [5], and virtual 3D environments. Both text-based presentations and illustrations are static in nature, and responses are typically entered via multiple choice without any direct time constraints.

In virtual 3D environments on the other hand, we see a greater variety of presentation modes and response mechanisms. Since they constitute the main mode of presentation in the studies presented in this book, we shall take a closer look at their variations and defining features. Note that the term virtual reality is often used interchangeably with virtual 3D environments, and further distinctions are made, for example between *desktop VR*, referring to environments displayed on a regular computer screen (e.g., [95, 96]), and *immersive VR* (e.g., [41, 104]). Immersive VR refers to the common understanding of VR, i.e., the presentation of the virtual world in a head-mounted display (HMD) with head-tracking, such as the Facebook Oculus Rift, or HTC

Vive HMDs. The key features of immersive VR are thus the participants' complete visual separation from the real world, and the first person view with head-tracking for a complete immersion into the virtual world. Finally, image projections on large screens or walls, encompassing a large part of the participants' visual field, as seen for example in [43], constitute somewhat of a middle ground between desktop and immersive VR. When it comes to the chosen perspective, most VR studies use a first person view [95, 41, 104], even though 3D environments allow the changing of perspectives, for example to that of a bystander at the scene, often with little additional programming effort.

With regards to input methods, VR offers various options. For example, the studies presented in this book, as well as [11, 39, 104], used keyboard inputs to choose between lanes or tracks. Joysticks, as used in [95, 96, 41, 42] present a more haptic, and possibly more 'personal' experience of interacting with the virtual world. Another option for actions such as pulling a lever in immersive VR would be to use hand-held VR controllers, which are tracked just like the HMD. However, the additional effort required to integrate further tracking devices may be hindering widespread use in behavioral ethics so far. Beyond multi-purpose input devices, driving simulators typically come with steering wheels and pedals, if not a complete car interior to maximize realism and immersion for car driving. While the analogue input of the steering wheel and pedals in driving simulators would certainly be beneficial for the feeling of actually driving a vehicle, it is not clear how one could achieve a binary, or otherwise discrete, response mechanism. For example, when presented with a choice between two lanes, each of which is blocked by an obstacle, analogue inputs and full control of the vehicle would likely result in many participants hitting the brakes or trying to carefully maneuver around the obstacles, thus avoid consciously making any difficult decisions. [43] approached this issue by stopping the time in the simulated environment before impact to present the participants with a multiple choice decision. However, doing so may reduce the participants' feeling of agency in the situation, which may or may not be a desired effect.

Besides various options of presenting the scenarios and assessing participants' responses, two important aspects distinguish simulated 3D environments from illustrations and text-based scenarios: The simulations naturally entail the passage of time, and they typically provide richer contextual information than vignettes or illustrations. The natural passage of time in 3D simulations increases the participants' immersion into the virtual world, and gives researchers the option of portraying sequences of events, or display movement relevant to the scenario.

However, it also limits response time windows in trolley-like scenarios, unless the action is frozen or slowed down significantly before the decision is made. This, however, may risk breaking the immersion to an extent. While event sequences and relevant movements in the scenario are one part of the rich contextual information, visual detail is the other. In detailed 3D environments, visual queues could, for example, be used to assess biases without making the participants explicitly aware of them, or can be used to imply the severity of a choice, without defining it precisely. The flip side of this is that it is often required to make intended consequences of the decision very obvious. [96], for example, opted to show the ensuing accidents graphically. Typically, the visual style of the 3D environments is aimed at realism [96, 39, 11], but researchers may also choose to go with simplified virtual environments [74].

While all of these choices essentially represent different ways of asking the participants how they would, or, how one ought to behave in a given situation, they are by no means equivalent per se. As noted in the previous section, research has shown that the framing of a dilemma in text-based settings can have a significant influence on response tendencies [56, 19]. Consider the finding of [26] that using a language other than the speaker's first language in text-based dilemma presentations leads to an increased tendency of utilitarian responses. The authors explain this by a greater psychological distance to the supposed events. Extending this logic to comparisons between different modes of presentation, one might expect that different levels of abstraction place the situation on different levels of emotional proximity to the participants, and similarly result in different response patterns. As a consequence, different modes of presentation may well also have an influence on the decision making process in traffic dilemmas. Indeed, studies have shown large discrepancies between different modes of assessment, raising the question whether moral judgment and moral actions should be treated as entirely distinct constructs [96, 41]. However, the dilemma scenarios used in the afore-mentioned studies emphasize the aforementioned clash of different moral schools of thought, while traffic dilemmas are usually aimed at the participants' evaluation of the potential victims and the environmental circumstances. Furthermore, the context of traffic scenarios is arguably closer to most people's day-to-day reality, possibly making it easier to fall back on existing evaluations or behavioral instincts. To what degree the methodology of assessment has an influence on the decision patterns in traffic dilemmas, thus, is one of the focus points of the studies presented in this book.

2.5 Behavioral Modelling

Behavioral modeling describes the construction, training, and inference of mathematical models of behavior. In our use case, the modeling objective would be the behavior of drivers in critical situations. The purpose for such models is typically to gain an understanding of underlying factors in the decision making process. With the development of self-driving cars, however, the field of behavioral ethics is no longer only driven by an academic interest in how we as humans perform moral judgment. Automated driving presents a direct link between academic research, philosophy, and a technology that will soon be ubiquitous in our everyday lives. It creates the challenge of having to translate ethical values into mathematical expressions, such that they can be used in an algorithm for decision-making in the real world. How to model ethical behavior is therefore of key practical interest for researchers, car manufacturers, and law-makers alike. As we will elaborate in more detail in the discussion section, behavioral models can both describe observations made in studies, and be used as decision making algorithms in automated vehicles. And while model architectures may differ between the two use-cases, a transfer of parameters, or matching of outcomes may be technically feasible.

In the literature, regression models (in this context also referred to as value-of-life models[1]) have been suggested as a possible solution for AVs in cases where higher-level rules fail to provide the system with clear instructions [50]. In simple terms, the model would work as follows: Each object that could potentially be harmed or damaged in a collision is assigned some positive value, and the model sums up the value of all "damages", weighted by the probability that these damages occur. This is done for each possible path the car could take, and the path with the lowest associated cost is chosen. However, whether such models are able to capture the complexity of human ethical decision-making was not clear prior to the studies in this book.

An account of a value-of-life model that is focused on a person's age is given by [71]. The authors conducted a large-scale survey in which people had to indicate in several instances, on which of two road-safety-

---

[1]The term value-of-life model can be used to emphasize when a model considers features of potential victims explicitly and assigns a value to each of these features. This gives rise to a combined score for each of the involved people, essentially modelling the likelihood of saving them as a sum of their features' valuations. Mind that the modelling of individual features is not an inherent property of regression models, and any number of external factors can be considered by these models as well. Thus, not all regression models are value-of-life models, and vice versa, but most models discussed in this book are both.

improvement measures they would rather spend a given budget. The available measures differed with respect to the age and expected number of people that would be saved, as well as whether the ones saved would be pedestrians or car drivers. The authors used a standard probit regression model to fit the observed data, and found the number of saved life-years to be the most important factor in the subjects' decision, allowing for specific values of life to be assigned to each age group.

In the studies presented in this book, we chose logistic regression models, since they are natural choice to model binary decisions. These are closely related to the probit regression models used in [71], and thus also constitute a form of value-of-life models. We use these to model factors influencing moral decisions on the side of the potential victims, on the side of the participants, and on the side of the assessment methodology, while also exploring different options of modelling decisions within the class of logistic regression models.

2.6   Motivation and Research Objective

This book aims to advance vehicle automation with regards to technology, ethics, and regulation. The primary focus is the analysis of behavioral preferences of humans in dilemmatic traffic situations, as well as an investigation of the used assessment methodologies and modelling techniques. From this and other findings in the field, we derive the state-of-the-art and highlight important directions and open questions in research and regulation. The ultimate goal is to drive the field towards a solution for automated and algorithmic ethical decision making. The secondary focus is on advances in deep neural networks, with implications for sensory systems in self-driving vehicles.

It is important to note that algorithmic decision-making is by definition pre-defined. If we assume that (1) the algorithm is deterministic given a complete description of the situation in question, and (2) the mapping from descriptions of the situation to decisions is reasonably smooth, i.e., similar situations lead to similar decisions, then we can test the algorithm's decision in essentially any possible scenario prior to its roll-out, by providing the associated inputs synthetically. The responsibility thus lies with manufacturers and lawmakers to assert that the decisions the system makes are in line with ethical norms and principles. However, there is no one defined set of ethical norms and principles, as these can differ to great extent both between and within societies and cultures. And since it is to be expected that the same decision making logic will be running on a manufacturer's entire fleet, which can be millions of cars, any small unwanted or unacceptable bias or design issue is greatly amplified and can have harsh real-world consequences. Underdeveloped decision making systems may lead to unnecessary harm and fatalities, and may show problematic biases in the way they allocate risk to different groups of road users. In turn, this could lead to expensive lawsuits, public backlash and slowed adaptation of the technology. Solving algorithmic decision-making is, for these reasons, both important and challenging.

The assessment of ethical norms and values of the general public in context specific settings notably only provides us with one view of what is "right" in dilemmatic situations. Other angles to look at the issue are provided by moral theories and frameworks such as utilitarianism or deontologism. Moreover, philosophers, scientists and politicians can weigh in with expert opinions. And finally, any solution to the issue must, of course, be compatible at least with the basic law and constitution of the country in which it is used.

One might wonder, then, if the solution has to be in line with existing law, why can't we simply apply today's traffic regulations to AVs in the same way we apply them to human drivers? As we will see in greater detail in publication four, current traffic regulations are generally not sufficient to provide a framework for the regulation or implementation of automated ethical decision making, or automated driving. This is owed to the fact that they were created for human road users, with all that this entails. Unlike humans, a pre-defined logic can be regulated very precisely, but it lacks a notion of common sense. It therefore requires precise instructions, e.g., as to under which circumstances regulations may or should be violated. This is a demanding challenge for the ethical and law community that may require laws for automation to accommodate a more mathematical language, rather than the more abstract human language. This notion is also part of the motivation behind our work on mathematical models of ethical decision making.

Our work on assessment methodologies is motivated by the aforementioned findings of framing effects in behavioral ethics. We are, therefore, not restricting the scope of our work to the question who should live and who should die when it comes to it, but strive to look at decision making in a nuanced way that takes into account all the relevant factors that impact our decisions, allowing us to draw conclusions about existing biases, and come to a solid picture of how we as humans make decisions in dilemmatic situations in road traffic.

At the outset of the work in this book, existing work on mathematical or algorithmic models of ethical behavior was somewhat underdeveloped. Similarly, differences with regards to the assessment methodology had hardly been explored, and little was known about the extent to which features of potential victims, such as age or sex, influence the decisions we deem right in this context. We thus set out to create a first point of reference in publication one, and overall aimed to answer the following four questions over the course of this book:

- Delivering a frame of reference: How do we as humans behave in traffic dilemma situations?

- How do we best assess human decisions? What biases are introduced by the assessment?

- How can we model human behavior in traffic dilemmas, and how do these models relate to decision making in AVs?

- What challenges remain on the road to automated ethical decision making in AVs?

Chapter 3

Using Virtual Reality to Assess Ethical Decisions in Road Traffic Scenarios: Applicability of Value-of-Life-Based Models and Influences of Time Pressure

Contributions

Frontiers in Behavioral Neuroscience: Leon René Sütfeld, Richard Gast, Peter König, and Gordon Pipa 2017

LRS: Conceptualization, Data curation, Formal analysis, Investigation, Methodology, Project administration, Visualization, Writing – original draft; RG: Data curation, Formal analysis, Writing – review & editing; PK: Supervision, Writing – review & editing; GP: Resources, Supervision, Writing – review & editing

Layman's Summary

With the advent of vehicle automation, one of the most fiercely debated questions became how an AV should behave in dilemma situations, i.e., when a decision has to be made between two (highly) undesirable outcomes. We approached this question by looking at the decisions humans make in such situations. To this end, we opted to observe human ethical decisions in a setting that was as close to the real world as practically feasible. Immersive virtual reality uses a head-mounted display (HMD) to virtually place the user inside a rendered 3D environment, tracking their head movement and matching it with the in-game camera, thereby creating the illusion of being inside the virtual world. We created a virtual world, as well as an application running the behavioral assessment within it specifically for the purpose of this study.

The study was designed to provide insight into how we as humans decide in traffic dilemmas. For instance, is the perceived value of obstacles approximately additive when combining obstacles in a group, and can we make out effects of age and gender? Beyond these questions, we analyzed what mathematical models would be appropriate to describe the decisions we make in these situations, and looked into the effects of time pressure.

Our results in this study outlined what subsequent studies ended up confirming: The decision making process can be well described by a value-of-life trade-off between the two options, and linear models are sufficiently complex to model this process. Increased time pressure drastically decreased consistency in the decisions, and we saw effects of group size and age of the potential victims on the decisions. This study laid the groundwork for future studies, as it demonstrated the the viability of immersive VR as an assessment environment for behavior in ethically challenging situations.

3.1 Introduction

Privately owned cars with autopilots first became a reality with a software update which Tesla Motors released to its fleet in October 2015, and many comparable systems will be on the market soon. While initially, these systems are likely to be restricted to highway use, they will eventually make their way into cities, with estimates predicting autonomous vehicles (AVs) dominating road traffic by the 2040s [91, 87]. The new technology is expected to reduce the number of car accidents significantly: The German Federal Statistics Agency reports that in 2015, 67% of all accidents with injuries to people were caused by driver misconduct. A 2008 survey by the National Highway Traffic Safety Administration (NHTSA) even showed that human error played a crucial role in 93% of traffic accidents in the US. These numbers outline the enormous potential of self-driving cars regarding road safety. In fact, [70] claim that self-driving cars will adjust their driving style and speed such that safe handling of any unexpected event is guaranteed at all times. However, this approach appears unrealistic for many mixed traffic (human and AVs) and inner city scenarios. To ensure absolute safety even in very unlikely events, the car would have to drive in an overly cautious manner, and as a result may be switched off by many drivers, or tempt other drivers to engage in risky overtaking. Other rare events, such as a distracted human driver swerving into the opposite lane, seem very hard to evade altogether under any circumstances. Even when completely taking human drivers out of the equation, we are left with a considerable number of accidents, caused, for instance, by technical or engineering failure [51]. Altogether, with over a billion cars in operation worldwide, the sheer amount of traffic virtually guarantees that, in spite of the overall expected reduction of accidents, critical situations will occur on a daily basis.

With accidents involving autonomous cars being and becoming a reality, ethical considerations will inevitably come into play. Any decision that involves risk of harm to a human or even an animal is considered to be an ethical decision. This includes everyday decisions, e.g., deciding if and when to take a minor risk in overtaking a cyclist. But it also includes quite rare situations in which a collision is unavoidable, but a decision can be made as to which obstacle to collide with. By relying on algorithms to make these decisions, a self-driving car's ethics come predefined by the engineer, whether it's done with sophisticated ethical systems or simple rules such as "always stay in the lane". This development poses a new challenge to the way we handle ethics. If human drivers are in an accident and make a bad decision from an ethical standpoint, we count in their favor that they had incomplete knowledge of the situation and only fractions of a second to make a decision. Therefore, we typically refrain from assigning any blame to them, morally or legally [48]. Algorithms in self-driving cars, on the other hand, can estimate the potential outcome of various options within milliseconds, and make a decision that factors in an extensive body of research, debates, and legislations [85]. Moreover, the same algorithms are likely to be used in thousands or millions of cars at a time. Situations that are highly unlikely for an individual car become highly probable over the whole fleet. This enhances the importance of getting it right, and unpreparedness to handle this type of situation may result in a significant number of bad decisions overall.

Ultimately, moral decision-making systems should be considered a necessity for self-driving cars [50]. The present study addresses the question of how to assess and how to model human moral decision-making in situations in which a collision is unavoidable and a decision has to be made as to which obstacle to collide with. We conducted a virtual reality (VR) study in which participants had to make exactly this type of decision for varying combinations of obstacles, and used the obtained data to train and evaluate a number of different ethical decision-making models. In the next section, we will review the current state of psychological research with respect to moral judgment and decision-making, and derive the outline for the present experiment.

3.1.1 The psychology of moral judgment

The scenario in this study can be seen as an adaptation of the trolley dilemma, a thought experiment commonly used in research on moral judgment and decision-making, in which a runaway trolley is heading toward a group of five people. The only way to save these five is to pull a lever and divert the trolley onto a different track, killing a single person instead [117]. The utilitarian choice here is to pull the lever and sacrifice one person in order to save five. By contrast, deontologism focuses on the rights of individuals, often putting these ahead of utilitarian considerations. From this perspective, the act of killing a person would be considered morally wrong, even if it means saving five other lives. In a popular alteration of the trolley problem, called the footbridge dilemma, the subject finds themself on a bridge over the tracks with a stranger. Pushing the stranger off the bridge in front of the oncoming train would stop the train and save the five people standing on the track. Interestingly, most people say they would pull the lever in the original trolley dilemma, but only a minority also says they would push the stranger off the bridge in the footbridge dilemma [56]. An extensive body of psychological research is concerned with the affective, cognitive and social mechanisms underlying this judgment, our ethical intuitions and behavior [68, 29, 124, 21, 4]. Most prominently, the dual process theory, put forward by [57], proposes two distinct cognitive systems in competition. The first is an intuitive, emotionally rooted system, eliciting negative affect when behavioral rules are violated, favoring decisions in line with deontological ethics. The second one is a controlled, reasoning-based system, favoring decisions corresponding with utilitarian ethics. Greene's dual-process theory thus explains the different endorsement rates of utilitarian behavior in the trolley and footbridge dilemma by the more emotionally engaging nature of the latter. Pushing a stranger off a bridge instead of pulling a lever requires personal force and uses harm as a means to an end, rather than as a side effect, both increasing the emotionality of the situation, and thus shifting focus to the system favoring deontological ethics [56]. Similarly, framing a dilemma as more personal ("I would do..." instead of "it is acceptable to...") and increasing the emotional proximity to the potential victim will also result in fewer utilitarian choices [56, 115]. Neuroscientific evidence is provided by [114], showing that disrupting the right dorsolateral prefrontal cortex, associated with emotional processing, increases the likelihood of utilitarian responses. [121] found an increased probability of utilitarian responses when inducing positive affect, arguing that the positive affect may cancel out the negative affect connected to rule violations.

However, the dual-process theory, based on the emotion-cognition distinction is not undisputed. [28] argues that while a distinction between competing processes is necessary, the distinction between affective and nonaffective processing is inadequate, since both processes must involve cognition, as well as affective content. Instead, he proposes a distinction based on two cognitive mechanisms borrowed from the field of reinforcement learning. The first is an action-based system, assigning reward values to possible actions in a given situation. These reward values are learned from experience and statically assigned to a given situation-action-pair. The second mechanism is outcome-based and relies on an underlying world model. In simplified terms, it predicts the consequences of the possible actions in a given situation and reassigns the value of the consequence to the action that leads to it. In the trolley dilemma, the outcome-based system would favor utilitarian behavior, and the action-based system would not intervene because the action of pulling a lever is not generally associated with negative reward. Conversely, the action of pushing a person off a bridge is associated with negative reward, thus explaining the lower endorsement rates of utilitarian behavior in the footbridge dilemma. Further evidence in favor of the action vs. outcome distinction in dual-process models is given, e.g., in [30] and [41].

In another theory in the realm of moral judgment, [62] aim to explain different views of opposing political camps (liberals and conservatives) with a model of morality based on five factors, and the relative importance of each of these factors to members of the political camps. Finally, based on a large body of neuroscientific evidence, [93] propose a detailed account of moral emotions as the foundation of our moral judgment. While none of the two entail concrete predictions with respect to moral decision-making in the trolley dilemma and similar scenarios, they demonstrate that the scope of the dual-process theories is limited, and that we are a long way from a comprehensive theory about the cognitive mechanisms governing our moral judgment and behavior.

3.1.2 Virtual reality assessment and effects of time contraints

While most of the aforementioned research relies on abstract, text-based presentations of dilemma situations, a growing number of studies makes use of the possibilities provided by virtual reality (VR) technology. VR, and in particular immersive VR using head-mounted displays (HMDs) and head-tracking, allows assessing moral behavior in a naturalistic way, immersing the subject in the situation, providing much richer contextual information, and allowing for more physical input methods. In an immersive VR version of the trolley dilemma, [95] were able to confirm the utilitarian choice's approval rate of 90%, previously found in text-based studies. Further, they found a negative correlation between emotional arousal and utilitarian choices, in line with the predictions of the dual process theory. In contrast to this, [96] found both emotional arousal and endorsements of utilitarian choices to be higher in a desktop-VR setting with 3D graphics on a desktop screen than in a text-based setting. While hinting towards a possible distinction between moral judgment and behavior, the results also suggest that features other than emotional arousal play a major role in our moral judgment. The authors argue that the contextual saliency (including a depiction of the train running over the virtual humans) may have shifted the subjects' focus from the action itself towards the outcome of their decision. The tendency to favor utilitarian judgment would then fit Cushman's account of the dual-process theory. In a similar study by [41], participants were confronted with the footbridge dilemma, either in an immersive VR environment or in a text-based scenario. In the text-based condition, endorsement of the utilitarian choice was low at around 10%, in line with expectations based on previous assessments. In the VR condition, however, subjects opted to push the stranger off the bridge in up to 70% of the trials. These results are again in line with Cushman's account of the dual-process theory, and make a strong case for the notion of moral judgment and moral behavior being distinct constructs. In a different approach, [104] varied the standard design of the trolley dilemma in multiple ways. First, they virtually placed participants in the trolley's cockpit instead of a bystanders' perspective. Second, they designed the track to split into three and blocked the middle track with a stationary trolley, which had to be avoided. Participants were thus forced to choose between the outside tracks, precluding the deontological option of not intervening in the situation. Third, the subjects had to react within 2.5 seconds after the obstacles became visible. Finally, in addition to varying the number of people on the available tracks, the authors added a number of trials with only one person standing on either of the available tracks. These differed in gender, ethnicity, and whether the person was facing towards the trolley or away from it. Unsurprisingly, the group was saved in 96% of the the one-vs.-many trials. In the single obstacle trials, significant differences were only found in the gender condition (deciding between a man and a woman), with men being sacrificed in around 58% of the cases.

The natural passing of time is a feature inherent to VR studies of this kind. While in principle, it would be possible to pause time in the virtual world, doing so might break immersion and would likely lessen the ecological validity of the obtained results. The previously mentioned studies all imposed some time constraints, but no systematic variation of response time windows was performed. Nevertheless, the dual-process theories would predict time pressure to influence our moral judgment. The action-based system in Cushman's account of the dual-process theory is thought to be simple and quick, while the outcome-based system involves higher cognitive load and is ultimately slower. Greene's account of the dual-process theory suggests that in emotionally engaging dilemmas, the controlled cognitive system needs to override the initial emotional response before making a utilitarian judgment [54]. Indeed, increased cognitive load during decision time was shown to increase response times in personal dilemmas, when a utilitarian response was given [58]. [97] showed that moral judgments can be changed with persuasive arguments, but additional time for deliberation was required for the change to occur. To the best of our knowledge, so far only one study systematically varied the length of response time windows. In [108], participants were either restricted to give a response within 8 seconds, or they had to first deliberate for three minutes. For high-conflict scenarios, such as the footbridge dilemma, higher time pressure led to fewer utilitarian responses. A second experiment in the same study supports this finding. When no time limitations were given, but one group was instructed to respond intuitively, and the other group was instructed to deliberate before entering a reaction, the intuitive group's response times were a lot shorter than the deliberate group's, and they gave fewer utilitarian responses.

In conclusion, VR studies have shown the importance of contextual cues for our decision-making and provide intriguing evidence for a distinction of moral judgment and behavior. Moreover, time constraints, as an inherent feature to VR setups, have been recognized as a factor in our moral decision-making. There is evidence suggesting that longer deliberation may facilitate utilitarian decisions in certain complex scenarios, but we still lack a systematic analysis of the influence of time pressure on moral judgment.

3.1.3 Modeling of human moral behavior

An important criterion that an ethical decision-making system for self-driving cars or other applications of machine ethics should meet is to make decisions in line with those made by humans. While complex and nuanced ethical models capable of replicating our cognitive processes are out of reach to date, simpler models may deliver adequate approximations of human moral behavior, when the scope of the model is confined to a small and specific set of scenarios. Value-of-life-based models stand to reason as a possible solution for any situation in which a decision has to be made as to which one of two or more people, animals, other obstacles, or groups thereof to collide with. An account of a value-of-life model that is focused on a person's age is given by [71]. The authors conducted a large-scale survey in which people had to indicate in several instances, on which of two road-safety-improvement measures they would rather spend a given budget. The available measures differed with respect to the age and expected number of people that would be saved, as well as whether the ones saved would be pedestrians or car drivers. The authors used a standard probit regression model to fit the observed data, and found that not the number of saved lives, but rather the number of saved life-years to be the most important factor in the subjects' decision, allowing for specific values of life to be assigned to each age group. Beyond this, they found pedestrians to be valued higher than car drivers of the same age, indicating consent as a factor in the valuation. While discriminating between potential human crash victims based on age, or possibly gender, is unlikely to gain general public acceptance, [50] suggests using value-of-life scales in cases where higher-level rules fail to provide the system with clear instructions. Furthermore, if we take animals into account, value-of-life scales stand to reason as a way of dealing with vastly differing probabilities. When a decision has to be made between killing a dog with near certainty and taking a 5% risk of injuring a human, how should the algorithm decide? We don't seem to take much issue with assigning different values of life to different species, and a system favoring pets over game or birds might be acceptable in the public eye. While this makes the case for at least some form of value-of-life model, it remains to be seen to what extent such models are able to capture the complexity of human ethical decision-making.

3.1.4   Deriving and outlining the experiment

As discussed in previous sections, our moral judgment is highly dependent on a wide variety of contextual factors, and there is no ground truth in our ethical intuitions that holds irrespective of context. We thus argue that any implementation of an ethical decision-making system for a specific context should be based on human decisions made in the same context. To date, our limited understanding of the cognitive processes involved prevents us from constructing a comprehensive ethical model for use in critical real-world scenarios. In the context of self-driving cars, value-of-life scales stand to reason as simple models of human ethical behavior when a collision is unavoidable, and an evaluation of their applicability in this context is the main focus of this study.

To this end, participants were placed in the driver's seat of a virtual car heading down a road in a suburban setting. Immersive VR technology was used to achieve a maximum in perceived presence in the virtual world. A wide variety of different obstacles, both animate and inanimate, were randomly paired and presented in the two lanes ahead of the driver. Participants had to decide which of the two they would save, and which they would run over. Since prolonged sessions in immersive VR can cause nausea and discomfort, we opted for a pooled experimental design with short sessions of 9 trials per condition and participant. We thus pooled the trials of all participants, and used this data set to train three different logistic regression models to predict the lane choice for a given combination of obstacles. (1) The pairing model uses each possible pairing of obstacles as a predictor. Here, a given prediction reflects the frequency with which one obstacle was chosen over the other in the direct comparisons. Since an obstacle is not represented with a single numerical value, the pairing model is not a value-of-life model, but serves as a frame of reference. (2) The obstacle model as-

signs one coefficient to each obstacle and uses the obstacles' occurrences
as predictors. We interpret these as the obstacle's value of life. (3) The
cluster model uses only one coefficient per category of obstacles, as they
resulted from a clustering based on the frequency with which each obstacle was hit.

We compare the three different models to test a set of hypothe-
ses. Our first hypothesis was that a one-dimensional value-of-life-based
model (i.e., the obstacle model) fully captures the complexities of pairwise comparisons. The obstacle model should thus be as accurate as the
pairing model. This would mean that our ethical decisions can be described by a simple model in which each possible obstacle is represented
by a single value, and the decision which obstacle to save is based only
on these respective values. We further hypothesized (Hypothesis 2) that
within-category distinctions, for example, between humans of different
age, are an important factor in the decisions. Specifically, the obstacle
model should prove to be superior to the cluster model. Furthermore,
since a certain amount of time pressure is inherent to naturalistic representations of this scenario, we investigated its influence on the decisions by varying the time to respond in two steps, giving participants a
response window of either 1 or 4 seconds. We found four seconds of decision time to induce relatively little time pressure in the used scenario,
while one second still left a sufficient amount of time to comprehend
and react. We hypothesized (Hypothesis 3) that more errors would be
made under increased time pressure, and that ethical decisions would
thus be less consistent across subjects in these trials. The dual-process
theories would predict a higher endorsement of utilitarian choices with
more time to deliberate (i.e., in the slow condition). However, for comparisons of single obstacles, there is no clearly defined utilitarian choice.
If anything, basing the decision in human vs. human trials on a person's
expected years to live could be considered utilitarian, and is partly covered in Hypothesis 2. Moreover, the omission of a lane change, despite
running over the more valuable obstacle, could be interpreted as a deontological choice, but we didn't formulate any directional hypothesis
regarding this factor prior to the study.

3.2 Methods

The experiment ran in a 3D virtual reality application, implemented
with the Unity game engine, using the Oculus Rift DK2 as the head-
mounted display. The audio was played through Bose QC25 and
Sennheiser HD215 headphones throughout the experiment. The participants were sitting in the driver's seat of a virtual car heading down
a suburban road. Eventually, two obstacles, one on either lane, blocked

the car's path and the participants had to choose which of the two obstacles to hit. Using the arrow keys on the keyboard, the participants were
able to switch between the two lanes at all times, up to a point approximately 15 meters before impact. This way, we provided a high level of
agency, intended to closely resemble manual car driving, while making
sure the decision could not be avoided by zig-zagging in the middle of
the road or crashing the car before reaching the obstacles. We used
17 different obstacles from three different categories, i.e., inanimate
objects, animals, and humans as single obstacles, as well as composite
obstacles. An empty lane was additionally used as a control. For each
trial, two of the 17+1 obstacles were pseudo-randomly chosen and al-
located to the two lanes, as was the starting lane of the participant's
car. A wall of fog at a fixed distance from the participant's point of view
controlled the exact time of the obstacle's onset. We varied the length
of the reaction time window by varying the fog distance and car speed,
resulting in a window of four seconds for the slow, and one second
for the fast condition. To indicate how much time was left to make a
decision at each point in time, a low-to-high sweep sound was played
as an acoustic cue. The sound started and ended on the same respective
frequencies in both conditions, thus sweeping through the frequency
band quicker in the fast condition. After the decision time window had
ended around 15 meters before impact, the car kept moving, completing any last instant lane changes. Right before impact, all movement
was frozen, all sounds stopped, and the screen faded to black, marking
the end of a trial. Figure 3.1 illustrates the chronological progression
of the trials in the fast and slow condition, and gives an overview of all
obstacles. The fast and slow trials were presented in separate blocks of
9 trials each. Two more blocks of trials were presented but not analyzed
in the current study, and all four blocks were presented in randomized
order. We chose obstacle pairings such that each obstacle typically appeared once per subject and block. The frequency of appearance of all
153 possible pairings, as well as the lane allocations and starting lanes,
were balanced over all subjects.

3.2.1 Sample and timeline

Our sample consists of 105 participants (76 male, 29 female) between
the age of 18 and 60 (mean: 31) years. We excluded one subject who
reported a partial misunderstanding of the task, as well as three outliers
whose decisions were the opposite of the model prediction (see below)
in more than 50% of their respective trials. Most of the participants
were university students or visitors of the GAP9 philosophy conference.

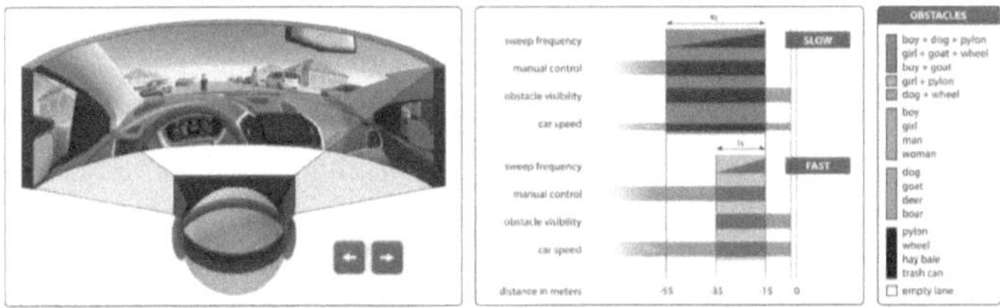

Figure 3.1: Left: Overview of the experimental setting. Middle: Timelines of the slow and fast conditions. Right: Overview of all obstacles used. Colors indicate cluster assignments.

Before participating, we informed all subjects about the study, potential risks and the option to abort the experiment at any time. They were also informed that the external screens would be turned off during the experiment, so that others could not observe their decisions. After signing a consent form, they were asked to put on the HMD and headphones, and then received all further information within the application. As a first task, they had to complete a training trial, avoiding three pylons by alternating between the lanes. Upon hitting a pylon, the training trial was repeated until completed without error. This procedure gave participants a chance to familiarize themselves with the VR environment, and it made sure they had understood the controls before entering the experimental trials. The study conformed to the Code of Ethics of the American Psychological Association, as well as to national guidelines, and was approved by the Osnabrück University's ethics committee.

3.3   Results

We pooled all data and did not consider between-subject differences in the analysis. In the experimental trials, the mean number of lane switches per trial was 0.816 in the slow and 1.037 in the fast condition. We estimated error rates for both conditions, using trials in which one of the lanes was empty. Hitting the only obstacle in such a trial was considered an error, as we find it safe to assume that the outcome in these trials is a result of inadvertently pressing the wrong button, instead of a meaningful decision. This event occurred in 2.8% of all trials containing an empty lane in the slow condition, and in 12.0% of the relevant trials in the fast condition. As a frame of reference, the chance level for this was at 50%.

3.3.1 Behavioral models

All models used in the present study were logistic regression models, using the occurrence of obstacle pairings, individual obstacles or clusters, i.e., obstacle categories (see below) in a particular trial as predictors for the lane choice. Furthermore, we added a constant offset and the trial's starting lane as predictors to all models. The constant offset allowed us to detect potential biases in the overall lane preference (left or right). Such a bias could occur, for example, when participants are used to right-hand traffic and feel that using the right-hand lane is generally more acceptable. Including the starting lane as a predictor allowed us to detect a bias to stay in the respective trials' starting lane – we would label this an omission bias – or to move away from the starting lane, i.e., a panic reaction bias.

A model's predicted probability of choosing to drive in the left lane is given by $p(Y = \text{left}|X) = \frac{1}{1+\exp(-X)}$, with X being the model-specific representation of a particular trial.

In the pairing model, a trial is represented as $X_p = c\omega_i + s\omega_s + \omega_b$, where ω_i is the coefficient for obstacle pairing i (e.g., [boy, woman]), $c \in [-1, 1]$ is the lane configuration in the respective trial (e.g., 1 if the boy is in the left lane, -1 if the woman is in the left lane), ω_s is the starting lane coefficient, $s \in [-1, 1]$ is the starting lane (1 if the starting lane is left, -1 is the starting lane is right), and ω_b is the coefficient for the lane bias. The model is thus making a prediction based on a general preference for one of the lanes, the starting lane of the respective trial, and which of the 153 possible pairings is presented in the trial, resulting in 155 parameters in total. Figure 3.2 left shows the predictions of the pairing model. Since each pairing of obstacles has its own free parameter, the model allows for intransitive and other complex relations between the obstacles. For example, in the slow condition, the pairing model deems the goat to be more valuable than the boar, and the boar to be more valuable than the deer, but the goat to be *less* valuable than the deer. Consequentially, there is no single value of life for an obstacle in the pairing model. An obstacle's value is only defined relative to each of the other obstacles.

In the obstacle model, a trial is represented as $X_o = \omega_{ro} - \omega_{lo} + s\omega_s + \omega_b$, with ω_{ro} and ω_{lo} being the coefficients for the right and left obstacle in the respective trial. Each obstacle is thus represented by a single characteristic value or value of life. All pairwise comparisons result directly from a subtraction of the respective two values of life. Thus, when sorting all obstacles according to their value of life on the abscissa and ordinate, the order in the vertical and horizontal direction is strictly monotonous (Figure 3.2, middle). Since there are 18 individual obstacles, the model has a total of 20 parameters, including the lane bias and starting lane coefficients.

Similarly, in the cluster model, a trial is represented as $X_c = \omega_{rc} - \omega_{lc} + s\omega_s + \omega_b$, with ω_{rc} and ω_{lc} being the coefficients of the clusters that the obstacles are assigned to. We performed bottom-up clustering and subsequent model selection to derive the ideal number of clusters and cluster allocations of all presented obstacles for the cluster model (see Figure 3.3). Logistic regression models were first constructed and fitted for all possible numbers of clusters, ranging from 17 to 1. We then performed the model comparison via the Bayesian Information Criterion (BIC). In the slow condition, the five clusters model was found to be the best of the cluster models. Notably, its cluster allocations are perfectly in line with a categorization in none, inanimate objects, animals, humans, and groups of humans and animals. In the fast condition, a four cluster solution was found to be ideal, and its cluster allocations don't align perfectly with the aforementioned semantic categorization. This is likely the result of the higher error rate in the fast condition. In order to still allow for a comparison between both conditions, we chose to use the aforementioned semantic categorization in five clusters for the fast condition, as well. For both conditions, the cluster model thus has five parameters for the obstacle clusters, resulting in a total of only seven parameters, including the lane bias and starting lane coefficients. Figure 3.2 right shows its predictions in the slow condition. The model uses only one free parameter per cluster of obstacles, resulting in a block structure. Since all obstacles within a cluster are considered equal in value of life, the difference in the value of life is always exactly zero for within-cluster comparisons. Those decisions, therefore, depend entirely on the starting lane and overall lane preference.

All models were fitted using the logistic regression algorithm in the scikit-learn (version 0.17.1) toolbox for Python. We set the regularization strength to a very low value of 10^{-9} and based the model selection on prediction accuracy via 10-fold cross-validation, as well as the Bayesian Information Criterion.

Pairing model vs. obstacle model

In a first step, we compare the pairing and the obstacle models. When modeling the training data set, models with a (much) higher number of free parameters can describe the data better. However, in cross-validation, potential overfitting can lead to a reduced performance of the more detailed model. Indeed, with a prediction accuracy of

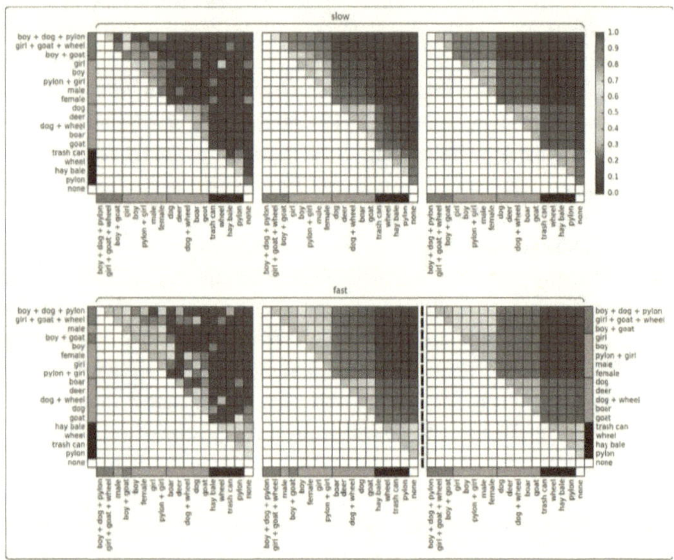

Figure 3.2: Model predictions. Top: slow condition, Bottom: fast condition, Left: pairing model, Middle: obstacle model, Right: cluster model. Colors indicate the probability of saving the row-obstacle (left) and sacrificing the column-obstacle (below). Pink, green, blue, and black bars indicate cluster assignments based on agglomerative clustering (see Figure 3.3). For means of comparability, the cluster model in the fast condition was fit with the semantic cluster assignments from the slow condition.

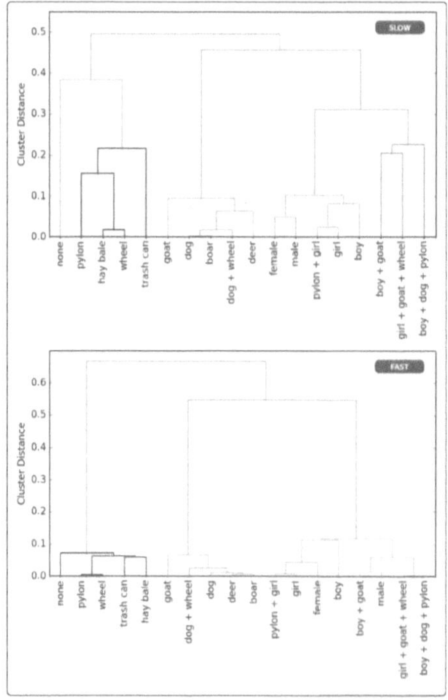

Figure 3.3: Dendrogram of bottom-up clustering, based on the observed frequencies with which each obstacle was spared (saved), for the slow and fast condition separately.

91.64% in the slow condition and 80.75% in the fast condition, the obstacle model is slightly superior to the detailed pairing model, with prediction accuracies of 89.33% and 78.77%, respectively. Despite our extensive data set with 909 trials per condition, the large number of parameters in the pairing model causes overfitting. This find translates to a much larger BIC value for the pairing model (see Table 3.1). Thus, our results strongly favor the obstacle model for its lower complexity and reduced risk of overfitting. This result, in combination with the high prediction accuracy of the obstacle model in the slow condition, confirms our first hypothesis, i.e., one-dimensional value-of-life-based models can adequately capture the ethical decisions we make in real life scenarios.

Obstacle model vs. cluster model

In the slow condition, the obstacle model's rankings of coefficient

| | slow | | | slow | slow | fast | fast |
model	SL	LB	parameters	BIC	accuracy	BIC	accuracy
pairing	x	x	155	1349.122	0.8933	1563.629	0.7877
obstacle	x	x	20	556.845	0.9164	770.827	0.8075
cluster	x	x	7	497.198	0.9120	691.389	0.8053
cluster		x	6	505.816	0.8922	685.797	0.8118
cluster	x		6	**491.852**	0.9120	684.656	0.8053
cluster			5	499.809	0.8636	**679.053**	0.8229

Table 3.1: BIC values and prediction accuracy based on 10-fold cross-validation for the three models in the slow condition. SL: including the starting lane as predictor, LB: including a constant offset as predictor to model a lane bias.

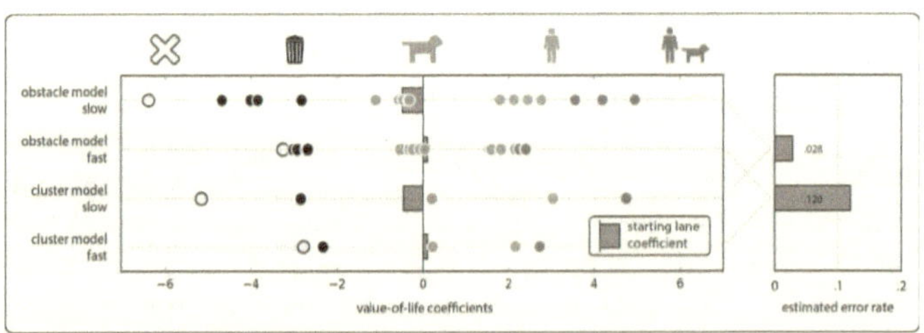

Figure 3.4: Left: Value-of-life coefficients by condition. Pictograms and colors indicate the categories empty lane, inanimate objects, animals, humans, and groups of humans and animals (left to right). Starting lane coefficients depicted as gray bars. Right: Relative frequency of 'saving' the empty lane object, used as error rate estimates, for fast and slow condition separately.

values within the categories mostly make sense, intuitively. For example, children are assigned higher values than adults (boy: 2.76, male adult: 2.12, corresponding to a 65.5% chance of saving the boy in a direct comparison with a male adult). Further, the dog is consistently found to be the most valuable of the animals. The prediction accuracies, however, are essentially even between the obstacle model (91.64%) and the cluster model (91.20%), with the cluster model scoring the lower BIC value, due to the reduced number of parameters (see Table 3.1). These findings are repeated in the fast condition. Prediction accuracies of the obstacle and cluster models are very close to each other (80.75% and 80.53%), and in terms of BIC values the cluster model is superior. We thus have to reject our second hypothesis, because the cluster model with five clusters is selected as superior to the obstacle model. In other words, we found no particular advantage of using obstacle-based predictors instead of category-based predictors.

Biases

To assess the two bias predictors' importance for the model, we ran another model comparison for three additional versions of the cluster model. All three additional versions were based on the above model, but the first variant dropped the starting lane predictor, the second variant dropped the lane bias, and the third variant dropped both predictors. In the slow condition, the cluster model omitting the lane bias, but including the starting lane predictor, scored the lowest BIC value of all models. Its prediction accuracy is the same as that of the previously assessed cluster model with both additional predictors (91.20%), making it the best explanatory model for the observations made in the slow condition (see Table 3.1). This is also reflected in the respective coefficients. When including the predictor of the lane bias, it was fit to a value of 0.15. The low value indicates only a very weak tendency to the left lane, which makes no significant contribution to the model fit. Thus, even in this rather realistic scenario, participants treated both lanes as equally valid driving lanes. The starting lane predictor was fitted to -0.47, indicating a reluctance to switch lanes in the face of a decision, constituting an omission bias. We can roughly quantify the extent of this reluctancy as being rather small, since coefficient differences between categories are in all cases magnitudes higher. The specific starting lane in a trial would therefore not affect the decisions in between category comparisons. It does, however, play a role in within-category decisions, as evidenced by the 4.8% gap in overall prediction accuracy between the cluster models with and without the starting lane as a predictor. In the fast condition, the best model, both in terms of prediction accuracy as well as BIC score, is the one omitting both bias predictors (see Table 3.1). By omitting the bias predictors, the prediction accuracy increases from 80.53% to 82.29%, exposing an overfit in the more complex model. In conclusion, the analysis of the bias predictors found lane preference to have no substantial influence on the decisions made in this paradigm, but did reveal an omission bias when facing similarly valued obstacles in the slow condition.

3.3.2 Influence of increased time pressure

We will now turn to a direct comparison of the slow and fast condition, to evaluate the effects of increased time pressure. The most notable difference between the two conditions is the (estimated) error rate of 12.0% in the fast condition, marking a four-fold increase from the slow condition. As for the cause of the errors, we would expect an increased omission bias if the errors were caused by a mere failure to react in time. Interestingly, this was not the case. Instead, we found an omission bias only in the slow and not in the fast condition, indicating that errors in the fast condition were equally a result of staying in and switching into the lane of the more valuable obstacle. A major increase in error rate also substantially decreases the expected prediction accuracy even for a perfect model. This is reflected in the prediction accuracies for models in the fast condition, which are on average roughly 10% below those for the corresponding models in the slow condition.

In the cluster analysis, we found a four cluster model to yield the lowest BIC values in the fast condition, instead of the five cluster model found to be ideal in the slow condition. Moreover, the cluster assignments for some of the obstacles are also different, and do no longer match the semantic categories perfectly (see Figure 3.2). These findings are consistent with the influence of increased noise in the data, and can therefore also be ascribed to the increased error rates. Since there is no matching cluster model for both the slow and fast condition, we included a comparison of the cluster models based on the semantically defined categorizations in Figure 3.4, but decided to focus on the obstacle model in the remainder of this comparison. In the obstacle model, the coefficient range in the fast condition was reduced to 50% of that in the slow condition (see Figure 3.4). Specifically, the obstacles on the extreme sides of the spectrum – the empty lane and the groups of humans and animals – aren't separated well from the adjacent obstacle categories. To statistically confirm this observed difference, we used a nested model approach with log-likelihood ratio tests. For the nested model, we fitted the joint dataset of fast and slow conditions to the obstacle model using 19 predictors, i.e., the 18 obstacles plus the starting lane. For the larger nesting model, we added a second set of

19 predictors. These 19 were duplicates of the first 19 predictors, but were fitted only on the slow condition trials. Together, these two sets formed a model with 38 predictors in total. The log-likelihood ratio test between the nesting and nested model was significant ($p = 0.037$), showing that the reduction in parameters between the two significantly reduces model accuracy. In other words, the difference between the two conditions is large enough to justify the use of two completely separate sets of parameters to describe them. This confirms our third hypothesis, i.e., increased time pressure significantly decreases the consistency in the answering patterns.

Another notable difference between the two conditions is that we no longer observe a bias towards sacrificing the male adults in direct standoffs with female adults. Instead, participants saved males in 4 out of 7 cases in the fast condition. The previously speculated tendency towards social desirability would likely rely on slower cognitive processes, and thus not come into effect in fast-paced intuitive decisions.

3.4 Discussion

We investigated the capability of logistic regression-based value-of-life models to predict human ethical decisions in road traffic scenarios with forced choice decisions, juxtaposing a variety of possible obstacles, including humans, animals, and inanimate objects. The analysis incorporated various contextual and psychological factors influencing our moral decision-making in these situations, and examined in particular the effects of severe time pressure.

Our first hypothesis was that a one-dimensional value-of-life-based model fully captures the complexities of pairwise comparisons. With prediction accuracies well above 90% in the slow condition, and clearly outperforming the more complex pairing model, the obstacle model proved to be capable of accurately predicting the moral decisions made in the pairwise comparisons. The first hypothesis was thus confirmed. Note that since we used a wide range of obstacles, we cannot preclude some more complex effects happening on a more detailed level. One possible example of such an effect is the following: In the slow condition, the obstacle model shows male and female adults to have comparable value-of-life coefficients with a slight advantage for the males (2.12 versus 1.79), predicting a 41.8% chance of sacrificing the male adult in a direct comparison. This prediction is based on all the trials it has seen, i.e., the full dataset including all possible combinations of the 18 obstacles. Still, adult males were actually sacrificed in 4 out of the 5 cases (80%) of direct comparisons between male and female adults. This observation is in line with [104], who also found males to be sacrificed more often in a direct comparison. Interestingly, the authors found the tendency to sacrifice males to be correlated with a general tendency to answer according to social desirability. In our study, the tendency to sacrifice males only pertains to the slow and not to the fast condition, which makes sense, if we assume that the effect is rooted in a tendency towards social desirability. Considerations of social desirability could be construed as part of the outcome-based system in Cushman's account of the dual-process theory, which is thought to be the slower one of the two processes. However, the low number of direct comparisons this figure is based on, and the exploratory nature of this find, dictate caution with respect to its interpretation. We consider it a leverage point for future research, but not a major result of this study.

Our second hypothesis was that within-category distinctions, for example between humans of different age, are an important factor in the decisions. This hypothesis could not be confirmed in this study, as the obstacle model failed to show an advantage over the cluster model in describing the collected data. However, there are hints at a meaningful structure within the clusters. For example, the obstacle model found children to have higher values than adults, and the dog, as the only common pet among the animals shown, to have the highest value within the animal cluster. Thus, given a larger data basis, we would still expect within-category distinctions to improve the predictions made by value-of-life models. In particular, we would expect age to play a role in human vs. human comparisons. Surveys by [27] and [71] have previously shown that the value we assign to someone's life decreases considerably with the person's age. To what degree these judgment-based findings would also be reflected in assessments of behavior is unclear, since judgment and behavior can yield dramatically different outcomes [96, 41]. Based on our findings, we speculate that the difference in value-of-life between people of different ages may be less pronounced in behavioral assessments, but more data is needed to clarify this point.

Irrespective of the exact outcome of such assessments, systems discriminating based on age, gender or other factors may be considered unacceptable by the public, as well as by lawmakers. Nevertheless, the idea of weighing lives against one another isn't generally rejected. As [13] showed, a majority of people would prefer a self-driving car acting in a utilitarian manner, at least when it isn't themselves, who are being sacrificed for the greater good. Independent of whether or not human lives should be weighed against one another, assigning different values of life to animals even seems to be the logical choice, judging from how differently we treat different species of animals in other aspects of life. Value-of-life models based on species would allow us to differentiate between common pets and other animals, and would give us a tool to deal with situations in which the death of an animal could be avoided by taking a minor risk of harm to a human.

Our third hypothesis was that ethical decisions would be less consistent across and within subjects when the time to react is reduced. This hypothesis was confirmed. The error rate was drastically increased, the cluster analysis revealed fewer clusters with slightly different cluster assignments, and the range of value-of-life coefficients was significantly reduced. However, we cannot deduct from our data whether the decisions made under time-pressure are in fact less clear-cut than decisions formed with more time for deliberation, or if the effect can be fully explained by the increased error rate. Still, a full second of time to react is a lot more than we typically encounter in real-life scenarios of this kind, and the weak consistency in the decision patterns is a sign that we are ill-equipped to make moral decisions quickly, even when the situation comes expectedly. We therefore argue that, under high time pressure, algorithmic decisions can be largely preferable to those made by humans.

Another noteworthy difference between the fast and slow condition concerns the omission bias, which we only found in the slow, but not in the fast condition. Participants were thus less likely to switch lanes and interfere in the situation when given more time to decide. This fact can be interpreted as a sign of a more deontological reasoning – choosing not to interfere in the situation, and possibly trying to reduce one's own guilt despite causing greater damage as a result. A tendency towards deontological reasoning with more time, however, conflicts with both Greene's and Cushman's accounts of the dual-process theory, as well as, e.g., [108], who found that more time to decide will cause a shift towards utilitarian responses. One possibly decisive difference between the present study, and most other studies touching on the aspect of time in moral decision-making, is the type of scenario used and the corresponding absolute response times. Typically, the scenarios used are relatively complex moral dilemmas, and response times lie in the 8-10 seconds range for short, and up to several minutes in the longer or unconstrained conditions [108, 58, 97]. In contrast, the reaction time windows of 4 and 1 seconds used in the present study rather represent a distinction between short deliberation and pure intuition. The fast condition may thus fall out of the dual-process theories scope.

In this study, we purposefully constructed a simple scenario with clearly defined outcomes, featuring the variables necessary to fit value-of-life models. With the general applicability of these value-of-life models established, a number of ensuing questions arise. For example, what influence a person's emotional and cognitive features have on their decision, how different probabilities of a collision or different expected crash severities affect our judgment, and how groups of multiple people or animals should be treated in such models. Moreover, the option of self-sacrifice has been prominently discussed in literature [86, 55, 48, 105], and was assessed via questionnaire in [13], but hasn't been included in behavioral studies so far. We speculate that immersion and perceived presence may have a particularly strong influence on decisions that touch upon one's own well-being. Beyond this, considerations of fairness need to be addressed as well – for example, if one person is standing on a sidewalk and another has carelessly stepped onto the street. While the choice of a wide range of obstacles has proven helpful in understanding the big picture, more research is needed to answer open questions about effects happening within the categories. The design choices we made allowed us to focus on the applicability of value-of-life models, but the present study does not provide a fleshed-out model for implementation in self-driving cars. Instead, it constitutes a starting point from which to investigate systematically, how a variety of other factors may influence our moral decisions in this type of scenario and how they could be implemented.

A limiting factor for this study is the use of only one instance of each of the presented obstacles. We tried to select and create 3D models that are as prototypical as possible for their respective classes, but we cannot rule out that the specific appearance of the obstacles may have had an impact on the decisions, and by extension, the coefficient values assigned to the obstacles. Future studies or assessments that put more emphasis on the interpretation of single value-of-life coefficients, should include a variety of instances of each obstacle. Furthermore, larger and explicitly balanced samples would be needed to obtain models sufficiently representative of a society's moral judgment. Another fair point of criticism concerns the plausibility of the presented scenario. There was no option of braking during up to 4 seconds of decision time, and the car was keyboard-controlled and could only perform full lane switches. While there were good reasons for these design choices, namely to allow for enough decision time and to enforce a clear decision based on an unambiguous scenario, they limit the virtual world's authenticity and may hinder the subjects' immersion. Unfortunately, this issue seems unavoidable in controlled experimental settings. We believe that the virtual world implemented for this study nevertheless fulfills a high standard of authenticity overall, and, under the given constraints, illustrates the scenarios in question as close to reality as currently possible.

Future studies should further investigate the role of the presentation mode in this specific context. We argue that based on moral dilemma studies, a distinction between judgment and behavior may be justified. However, it remains to be seen if there is a seizable difference for specifically the kind of situations used in this study that justifies the special effort that goes into the design of a virtual reality environment. Finally, based on our findings, the influence of time pressure could be assessed in greater detail, expanding the considered time frames beyond the 1-4 seconds range.

3.5 Conclusion

We argue that the high contextual dependency of moral decisions and the large number of ethically relevant decisions that self-driving cars will have to make, call for ethical models based on human decisions made in comparable situations. We showed that in the confined scope of unavoidable collisions in road traffic, simple value-of-life models approximate human moral decisions well. We thus argue that these models are a viable solution for real world applications in self-driving cars. With respect to trust in the public eye, their simplicity could constitute a key advantage over more sophisticated models, such as neural networks. Furthermore, regression models can include additional factors such as probabilities of injuries for the parties involved, and help to make reasonable decisions in situations where these differ greatly. They also provide an easy option to deal with a vast number of possible obstacles, by testing a few and making reasonable interpolations, e.g., for people of different age, taking away the requirement of assessing each conceivable obstacle individually. That being said, the modeling of within-cluster differences, e.g., between humans of different ages or between different species of animals, failed to improve upon the rather coarse cluster models. We further found time pressure, as an inherent feature to naturalistic portraits of the scenario in question, to considerably decrease the consistency in the decision patterns and call for more investigation of the effect of time pressure on moral decision-making. Overall, we argue that this line of research, despite being met with some skepticism [70], is important to manufacturers and lawmakers. The sheer expected number of incidents where moral judgment comes into play creates a necessity for ethical decision-making systems in self-driving cars [50]. We therefore hope to see more efforts towards establishing a sound basis for the methodology of empirically assessing human ethics in the future, as the topic is becoming increasingly important with more advances in technology.

Chapter 4

Response: Commentary: Using Virtual Reality to Assess Ethical Decisions in Road Traffic Scenarios: Applicability of Value-of-Life-Based Models and Influences of Time Pressure

Contributions

Frontiers in Behavioral Neuroscience: Leon René Sütfeld, Richard Gast, Peter König, and Gordon Pipa 2018

LRS: Writing – original draft; RG: Writing – review & editing; PK: Supervision, Writing – review & editing; GP: Supervision, Writing – review & editing

Layman's Summary

In a reaction to the previous paper, [75] brought up two objections to our approach. Firstly, he argued that we are implying the validity of the meta-ethical position of particularism, i.e., that there are no moral principles and the right action is determined solely by contextual factors. Secondly, he found that we are committing an is/ought-fallacy by suggesting that human snap-judgments show us the morally right behavior. In this publication, we respond to his objections regarding the assessment approach, before elaborating on the applicability of the chosen modeling approach. Finally, we lay out why empirical observations can deliver a valuable frame of reference for algorithmic decision making in this context.

4.1 Introduction

In the paper discussed [109], we examined the feasibility of using virtual reality (VR) as an assessment method for models of human moral behavior in road traffic scenarios. Furthermore, this experimental approach allowed us to analyze the applicability of logistic regression-based value-of-life models for modeling human behavior. We consider this study to be a contribution to the discussion about ethical decision-making systems in autonomous vehicles (AVs).

4.2 The Chosen Assessment Approach

In a recent commentary on this paper, [75] brings up two objections to the approach:

In the initial study, we cite evidence showing that human moral intuitions differ depending on a variety of contextual variables. Keeling argues that we use this to infer the validity of the meta-ethical position of particularism [31], and that this inference is not necessarily justified. He further contends that our "answer to the moral design problem depends on the plausibility of this inference."

In our paper, we argue on the level of moral behavior and moral intuitions, which we can experimentally assess and describe. From the evidence cited, we conclude that these are indeed highly dependent on contextual factors. We thus argue that in order to learn about our behavior and moral intuitions in a particular real-world scenario, it is reasonable to match the contextual factors of the assessment with those of the scenario in question, making the case for a VR assessment as a starting point for this line of research. The experimental data presented and the conclusions based on it are, therefore, not dependent on a specific position in the views on particularism vs. generalism, but independent of this controversy.

Keeling argues that we are committed to the claim that "the right thing to do in AV collisions is determined by facts about human snap-judgements", and that this is not a valid claim, since "[h]umans are sensitive to the pressures of a collision, and under this pressure, our critical thinking capacities break-down".

The used term "snap-judgements" refers to the severe time constraints we encounter in real-life situations of this kind, and it implies that the decisions are likely to differ qualitatively from more elaborated decisions. However, the study used two conditions differing in the degree of time pressure. They delivered qualitatively similar results, giving us no indication that the cognitive processes leading to decisions in such situations differ qualitatively within the time range investigated. Furthermore, for the longer of the two conditions (4s), we observe a surprisingly high amount of consistency, which is at odds with the idea of a "break-down of critical thinking capacities". If, or to what extent even longer decision time frames might impact the decision outcome remains to be investigated, but we can take the high consensus among the participants as an indicator that the decisions made were far from arbitrary.

Beyond this, we see time constraint as only one of several factors in the assessment methodology that can potentially play a role in our decisions. These factors, including the level of abstraction (e.g., text-based versus naturalistic), and the level of immersion in the presentation (e.g., immersive VR versus desktop VR), should be experimentally investigated to get a clearer picture of how stable our moral intuitions are across assessment modes, and to what extent the decision patterns in one assessment mode may give a valid and generalizable account of our moral intuitions for the setting in question.

4.3   Applicability of the Modeling Approach

With respect to what ought to be done and the real-world applicability of the approach suggested, we need to make a distinction between the empirical findings we report and the general approach used to model moral decisions.

First, probabilistic decision-making systems seem unavoidable for self-driving cars. Situations encountered in real life are almost never prototypical, and there is a large number of factors that can play a role in our moral assessment. To what degree and with which probability each of a large number of factors is present in a given situation varies on a continuous spectrum. Any categorical decision-making system will thus either fail to capture all possible combinations of circumstances, have arbitrary decision boundaries, or it will be too large to be fully comprehensible.

Second, we showed that value-of-life-based logistic regression models are generally suited to describe the empirically observed decisions of our subjects in this kind of dilemma situation. Using this class of models, however, does not mean we suggest to unreflectively copy human behavior for cars. Since the straightforward interpretability of the model parameters allows us to superimpose higher-level rules, initial model parameters, possibly obtained by an empirical assessment, could be modified, and isolated factors could be excluded from the model to make it comply with existing jurisdiction or normative theories.

4.4   Applicability of Empirical Observations

The question remains as to whether ethical decision-making systems based on the empirical assessment of moral intuitions are preferable to systems based on more elaborate and/or normative approaches. In this context, Keeling suggests reverting to "one of our best moral theories, such as utilitarianism or contractualism". However, describing a moral theory as "best" places it on a quantitative scale, questioning the normative character of the analysis. By proposing two competing moral theories, Keeling also brings up the issue that to this date, no agreement has been reached about which of the various moral theories is the right one. Moreover, normative theories typically come with considerable shortcomings for the purpose at hand. Utilitarianism, for instance, would suggest the sacrifice of innocent bystanders if it means reducing the overall harm. It would further propose colliding with the best protected opposing party (potentially punishing cyclists for wearing a helmet), and it lacks objective quantification of harm for the different options in a given situation. The latter point is particularly problematic, as it leaves us without concrete guidelines on how to behave in more complex situations. Therefore, we cannot simply revert to one of the established normative theories. Empirical observations, on the other hand, can deliver a frame of reference for higher-order considerations:

They can guide the decision between different normative theories in situations where these contradict each other. They can deliver initial (numerical) values for mathematical models used in AVs, i.e., they can guide the quantification of rules where the qualitative direction is provided by higher-order rules. They can highlight aspects that are underrepresented in normative moral theories, but may play a decisive role in our behavior (e.g., the value of animals). They can highlight where normative theories may be at odds with the moral intuitions in a society, and can thus motivate a re-evaluation of particular aspects of a particular normative theory. In conclusion, we argue that empirical observations play an important role in informing the debate, as well as in determining the rules for implementing moral decision-making systems for AVs.

Chapter 5

How does the method change what we measure? Comparing virtual reality and text-based surveys for the assessment of moral decisions in traffic dilemmas

Contributions

PLOS ONE: Leon René Sütfeld, Benedict V. Ehinger, Peter König, and Gordon Pipa 2019

LRS: Conceptualization, Data curation, Formal analysis, Investigation, Methodology, Project administration, Software, Visualization, Writing – original draft; BVE: Data curation, Formal analysis, Writing – original draft, Writing – review & editing; PK: Supervision, Writing – review & editing; GP: Resources, Supervision, Writing – review & editing

Layman's Summary

Based on the findings from the previous study, we set out to get a deeper insight into the dependencies between the methodology of the assessment and its outcome. Prior to this study, a growing number of publications had chosen very comparable forced-choice paradigms to investigate how we as humans behave in dilemma situations in road traffic. However, the task was provided in a range of different ways, from entirely text-based without any time pressure, to fully immersive VR experiences with only a very short response window to make a decision.

In this study we compared abstract, text-based questioning to image-based presentations of the task, as well as VR versus non-VR presentations of the same scenarios, and found the response patterns to be remarkably consistent across conditions. This is an indication that the underlying construct giving rise to the decisions is robust, and not easily manipulable by contextual cues. By extension, this supports the validity of various studies in this field, and gives more weight to their results. For example, while the very large number of participants in the Moral Machine study [5] makes for high reliability of the observed results, the study on its own is restricted to a single style of presentation. It could therefore be the case that various design choices within the study led to a systematic shift in the response pattern. Our results suggest that this is unlikely, or at least that any such shifts would be negligible compared to the main factors influencing the decisions.

5.1 Introduction

Ethical considerations concerning autonomous machines and, in particular, self-driving cars have recently gained widespread attention in research, media and society. Questions of trade-offs between utility and safety, liability in the case of accidents, and biases in the detection of ethnic minorities are just some of the open ethical issues brought up by the development of this technology [12, 72, 67]. Most prominently, the question how an automated vehicle (AV) should behave in an ethical dilemma situation has been addressed in a large number of publications [13, 109, 39, 11, 43, 83, 5]. In these, the problem is typically broken down into a series of forced choice decisions between two options, akin to the trolley dilemma [117], and the decision patterns of participants are analyzed to infer what factors play a role in their decision making, and to which degree. While the purpose of such studies is not to provide a blueprint for the behavior of self-driving cars, their findings can inform the debate, point out where our intuitive moral judgment is at odds with moral theories and regulations, and deliver initial numerical values for formal decision making models [110]. On the regulatory side, a first advance towards defining a legal framework for the use of AVs was undertaken by an ethics commission of the German Federal Ministry of Transport and Digital Infrastructure [88]. With respect to decisions in dilemma situations, the commission precludes the consideration of individual features such as age or gender, but remains inconclusive about the consideration of the number of people harmed in any given option. The report also doesn't offer concrete suggestions for the design, implementation, or regulation of ethical decision making systems, citing a need for more research. There is consensus, however, that the systems need to be transparent, and the rules they obey to be suitably communicated to ensure public acceptance of AVs.

While it remains debatable which factors should be taken into account to make a decision in dilemma situations, a frame of reference can be derived from human decision making. The MIT's Moral Machine is a large-scale survey analyzing various factors that influence our moral decisions in dilemma situations, distinguishing between features of the situation (termed global preferences), and individual variations between the participants [5]. In the analysis of the global preferences, the largest effects were found for favoring humans over pets, larger groups of characters over smaller ones, and younger people over older ones. Further notable effects were found favoring those who behave lawfully, those with a higher social status, the physically fit, females over males, and pedestrians over passengers of the AV. A small effect was also found favoring inaction over action, suggesting that some reluctance to interfere in such a situation is part of our moral intuitions, but is often outweighed by utilitarian considerations. On the side of individual variations among the participants, small effects were found, for instance, for the participants' age and gender. Many of these findings are qualitatively corroborated by other studies. For example, [39, 11, 83, 43] all found strong tendencies towards favoring larger groups and younger people, and [109] previously found strong effects towards favoring humans over animals. Thus, studies report a variety of factors that influence human decision making in dilemma situations.

Interestingly, we observed a large variety in the approaches used to assess the factors of human decision making. The Moral Machine, for example, is a web browser-based survey using simple birds-eye view drawings of the scenarios in question [5]. By contrast, [109, 39, 11] used interactive virtual reality (VR) applications, showing the scenarios from the driver's perspective. Here, the decisions had to be made in real time, with response time windows of four seconds in most cases. [43] placed participants in a driving simulator, also presenting the scenarios from an immersive first person view, but freezing the scene at decision time and supplying additional information about the situation using text-overlays. [83, 71, 13], on the other hand, used predominantly or entirely text-based surveys in their studies. The large differences between these approaches raises the question, to what extent the same underlying construct is measured. In fact, we know from studies in the field of empirical ethics that contextual factors and the way we frame the question can have a sizable impact on the ethical decisions we make [56, 121, 115]. The large discrepancies between moral decisions in VR and text-based assessments, found in [96, 41] even suggest that moral judgment and moral action may be distinct constructs. However, the thought experiments used in empirical ethics are usually constructed to emphasize a clash of different moral schools of thought – typically deontologism, focused on moral rules that must not be broken ("do not actively kill another person"), and utilitarianism, focused on minimizing overall harm, thus saving the largest amount of people. Unlike most classical dilemma thought experiments, traffic dilemmas are not typically designed to emphasize a clash of different moral intuitions. Instead, they are usually aimed at the participants' evaluation of the potential victims and the environmental circumstances. The context of traffic scenarios is also arguably closer to most people's day-to-day reality, possibly making it easier to fall back on existing evaluations or behavioral instincts. To what degree the methodology of assessment has an influence on the decision patterns in traffic dilemmas, thus, remains an open and very relevant question that we address in the present work.

Research objective

In this paper, we analyze how the participants' behavior and decisions are influenced by the presentation of traffic dilemmas. The approaches used in the literature often vary in multiple aspects, making it difficult to trace how these aspects influence the participants' behavior. These aspects of the assessment can, to a large extent, be broken down into the level of abstraction, the presentation modality (desktop vs. virtual reality), and varying degrees of time pressure. To isolate the corresponding effects, we designed two traffic dilemma studies, in which we systematically vary the presentation of the dilemmas across these three dimensions. All in all, the two studies cover a spectrum of presentation styles from text-based questionnaires to a fully immersive VR experience akin to [109]. The effects of time pressure on the decision making process were examined, since some form of time limitation is an inherent aspect of decision making in immersive VR. Ultimately, this establishes how the different approaches used in the literature relate to each other, and it can inform us about potential biases in the participants' moral judgment connected to the assessment methodology.

In our statistical models, we also included personal features of the participants as predictive factors of behavior. These include the participants' age and gender, as well as two more variables of particular interest in this context: Video game experience and social desirability. Video game experience was included as a potentially explanatory variable, since virtual reality studies are arguably similar to video games in terms of visual and acoustic presentation, as well as user input. Frequent video game players might, therefore, have a different perception of the stakes involved in their decisions. Social desirability, i.e., a tendency in participants to answer in accordance with social norms instead of their true beliefs, might lead to systematic shifts in the decision patterns, so we assessed this tendency with the social desirability scale (SDS-17) questionnaire [107], and incorporated the respective scores as factors in the analysis.

5.2 Methods

Study 1

In the first study, we employed a 2x2 experimental design with the factors level of abstraction (naturalistic vs. text-based; within subjects) and presentation modality (VR vs. 2D desktop monitor; between subjects). *Levels of abstraction:* The naturalistic settings featured a rendered 3D environment, showing the dilemma situation from the driver's first-person view (see Fig 5.1). By contrast, the text-based settings replaced the 3D environment with text and simple visual indicators on a gray background. *Modality:* In both the naturalistic VR and text-based VR settings, participants wore a head-mounted display (HTC Vive) and headphones, allowing them to freely look around the respective environment. They had to make their decision within 4.0s (naturalistic) and 4.4s (text-based), respectively. In the desktop modality, participants were presented with a fixed screen, and the time to make a decision was unlimited in both naturalistic and text-based conditions. For a more detailed description of the experimental conditions, please refer to S1 Appendix: Conditions and controls.

The four conditions spanned a spectrum from *questionnaire-like* (text-based desktop) to *as realistic as possible* (naturalistic VR), while allowing us to treat the level of abstraction and the modality as separate factors in the analysis. We recruited 88 participants, mostly from the local university, and had to exclude three due to misunderstanding the instructions or crashes of the application. The remaining 85 participants (43 females, 42 males, mean age 23.0, for further information see S1 Table: Sample overview) were randomly assigned to either the VR (43) or desktop (42) condition, and reported their age and gender in the application before the trials started. In the experiment, they were presented with a block of 20 trials in the naturalistic setting, then with a block of 20 trials in the text-based setting, or vice versa (order assigned randomly). In each trial, participants chose which of two single obstacles on the road ahead of them they would rather spare, with the obstacles being randomly drawn from a pool of animals (dog, goat, and boar), and humans of different gender and age (young, adult, or elderly). Additionally, some trials featured an empty lane, as a form of sanity check.

Study 2

In the second study, we again employed a 2x2 experimental design, this time with the factors level of abstraction (naturalistic vs. text-based; within subjects) and speed (slow vs. fast; within subjects). The slow condition was identical to study 1, the fast conditions had smaller response time windows of 1.2s (naturalistic) and 1.6s (text-based). Fast response times in the naturalistic setting were achieved by increasing the car's speed and decreasing the viewing distance. All conditions used

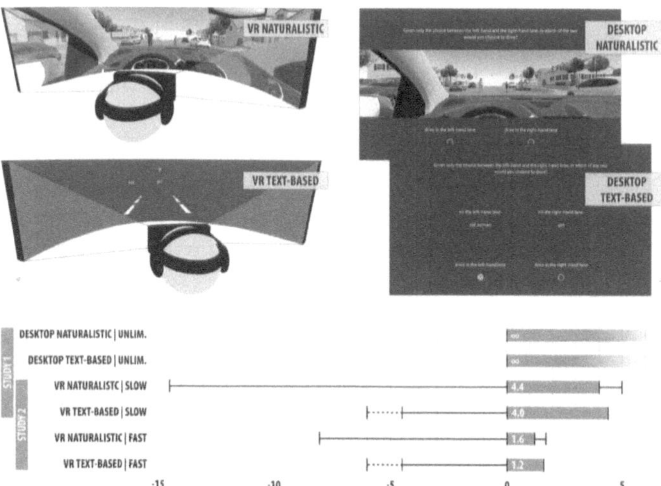

Figure 5.1: **Overview of the experimental conditions and time lines.** Black T-lines on the left indicate trial and control onsets, dashed lines indicate variable onset times. Blue bars indicate response time windows from visibility onset of the obstacles (VR: appearing from the fog) until car control offset, and black T-lines on the right indicate the time the car kept moving after control offset, i.e., the end of the trial. VR conditions featured an additional 1.5s of fade-to-black time (not indicated in the graphic).

the VR environment as described in study 1. The experimental procedure was divided into blocks by level of abstraction, showing at random either both naturalistic or both text-based conditions in a row. The order of the condition (speed) within these blocks was randomized. Each condition consisted of 7 trials, which were largely identical to those in the first study, except that animals were excluded from the obstacle pool to get a larger number of human vs. human trials. We recruited 107 participants, but had to exclude 14 due to an error in the application. Of the remaining 93 participants, 58 were females, and 35 males (mean age 21.3, for further information see S1 Table: Sample overview). Subjects reported their gender and age before the main experiment, filled in a short post-hoc questionnaire, as well as paper-based version of the SDS-17 questionnaire [107] (an assessment of their social desirability) after finishing the experimental trials.

Both studies conformed to the Code of Ethics of the American Psychological Association, as well as to national guidelines, and were approved by the Osnabrück University's ethics committee. A more detailed description of the conditions and controls can be found in S1 Appendix: Conditions and controls. The used hardware and the precise experimental timelines are defined in S2 Appendix: Timelines and hardware.

Statistical modelling

For the behavioral analysis, we employ Bayesian hierarchical logistic regression models to predict which lane a participant would choose, based on a number of explanatory variables characterizing the trial.

In its basic form, logistic regression models the probability of the outcome of a binary dependent variable. That is, the model finds the set of parameters that jointly determine the probability of finding a positive or negative outcome in a specific trial, maximizing the likelihood of observing the experimental data as they are. With some simplifications, one can interpret the parameters estimated by the model as modifiers of an object's ethical valuation. Positive log-odds thus indicate higher ethical valuation of the feature in question, negative log-odds a lower valuation.

In Bayesian statistics, we begin the modelling process with prior distributions, expressing our knowledge or belief about the impact of the modeled variables before seeing the data. In the model fitting step, the variables, or model parameters, are then approximated to realize a compromise between the chosen priors and the best fit for the observations made. The resulting distributions, called posterior distributions, ultimately represent the knowledge we have about the model parameters after seeing the data, with the posterior mean representing our best guess for the true impact of a given variable. In this analysis, we used weakly regularizing priors, representing a prior believe that the variables do not take on extreme values (see S4 Appendix: Bayesian logistic regression model specifications).

In this framework, no classical significance tests are performed. Instead, the evidence is treated as being on a continuous scale. The sign and magnitude of a parameter tell us about the direction and size of an effect, while the credibility intervals tell us how certain we can be that it is different from zero. Additionally, the Bayes factor provides a measure for how much our knowledge about a given parameter changed from the prior, based on the observations we made. Bayes factors between $\frac{1}{3}$ and 3 are generally regarded as inconclusive, with anything below $\frac{1}{3}$ being regarded as evidence against the null hypothesis (the hypothesis that the parameter has no influence on the outcome), and anything above 3 regarded evidence in favor of the null hypothesis.

The variables we used in the model can be divided into three categories: (1) Global preferences, such as age or gender of the potential victims, (2) features of the assessment, i.e., modality, abstraction and response time, and (3) individual features of the participants, such as their age or SDS-17 score. For all models, we used the maximal multi-level model [8], similar to the Bradley-Terry-Luce model of paired comparisons [14], with parameters for individual subjects on the second level. We fitted one model per study, and both models made use of the features of the portrayed situation and features of the assessment. In study two, the larger number of relevant trials allowed us to further include the individual subjects' features. Besides the main effects, we restricted the interactions considered in the model. For study 1, we modeled only interactions between global preferences, abstraction and modality. In study 2, we modeled interactions between global preferences, abstraction and speed, as well as between global preferences and each of the individual participant's factors. The model specifications and the chosen weakly informative priors are laid out in more detail in S4 Appendix: Bayesian logistic regression model specifications.

5.3 Results

The absolute rates of saving obstacles of different age groups and genders in both studies are provided in Fig 5.2. This descriptive view shows us that across conditions, females were saved in about 60% of all cases, children were saved in about 90% of all cases, and the elderly were saved in about 10% of all cases. Since the differences between the experimental conditions are small in comparison to these results, we can already infer that the potential victims' age and gender were dominant factors in the participants' decisions.

5.3.1 Global preferences

For a detailed assessment of all involved factors in the decision making process, we used a Bayesian logistic regression model analysis. We refer to features of the potential victims, such as their gender and age, as well as the lane they are in, and the lane the participants' car is in at the onset of the obstacle, as global preferences. The magnitude of influence these features have on the outcome of a trial for both studies are shown in Fig 5.3, and corresponding tables can be found in S5 Appendix: Regression coefficients. *Lane bias*: A lane bias describes a tendency to prefer either the right or the left lane, irrespective of the obstacles in those lanes. The mean a posteriori log-odds estimates for this are very close to 0, with the posterior distribution carrying a lot of weight on

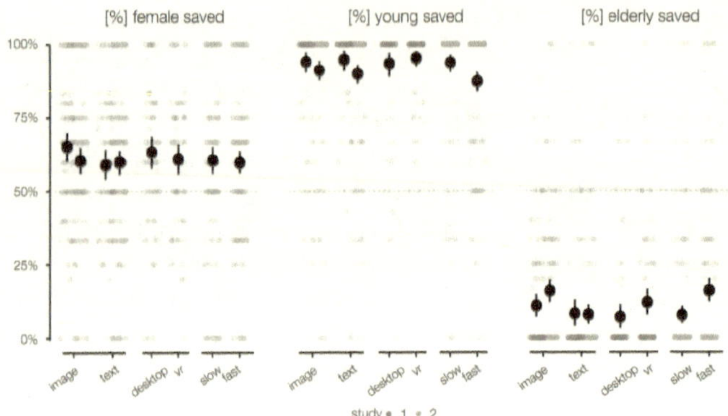

Figure 5.2: **Saving rates.** Rates of saving young (left), male (middle) and elderly (right) people, by study and experimental condition.

either side of it and the Bayes factors strongly preferring no bias (study 1: -0.2, CI_{95}: $-0.5, 0.2$, BF_{H0}: 11.1; study 2: 0.1, CI_{95}: $-0.2, 0.3$, BF_{H0}: 21.3). Thus, while it would be plausible that a right-hand lane bias exists due to right-hand traffic in Germany, any general lane bias in this sample is minimal at best and would not have any notable effect on their decisions. *Omission bias*: An omission bias shows a general tendency towards inaction, which may be rooted in an aversion to active causation of harm. This bias was observed in study 1 (1.4, CI_{95}: 0.4, 2.6, BF_{H0}: 0.1) and inconclusive in study 2 (0.6, CI_{95}: $-0.1, 1.3$, BF_{H0}: 1.7). However, even in study 1 its size is small in comparison to the effects of gender and age, and played only a subordinate role in the participants' decisions. *Gender bias*: A considerable bias in favor of female obstacles was observed in both studies (study 1: 3.0, CI_{95}: 2.3, 3.8, BF_{H0}: 0.0; study 2: 2.6, CI_{95}: 1.9, 3.4, BF_{H0}: 0.0). The small random effects standard deviations for the estimated individual parameters (study 1: 0.7, CI_{95}: 0.03, 1.87, study 2: 0.45, CI_{95}:0.97, 2.73) also indicate a high consensus within the sample population. *Age bias*: The age of the potential victims had the largest influence of all considered factors on the trials' outcomes, with mean posterior estimates of 9.5 (young) and -7.4 (elderly) in study 1, and 8.1 (young) and -6.9 (elderly) in study 2. Interestingly, the between-subjects variance of the age bias is fairly large, indicating a weaker consensus about the extent of the age bias in the sample population (study 1: 3.9, CI_{95}: 0.1, 7.4, study 2: 3.4, CI_{95}: 2.1, 5.0). Overall, these findings are in line with the existing literature [109, 11, 5], supporting the general suitability of this paradigm to test influences of the assessment methodology on the participants' behavior.

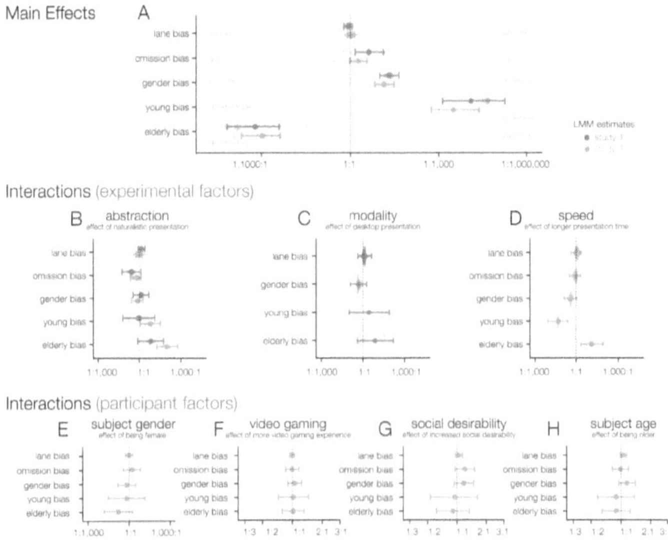

Figure 5.3: **Bayesian hierarchical logistic regression results.** A) Main effects of a logistic linear mixed regression model for both studies. Error bars indicate 95% credible intervals around the median posterior estimate of the group estimate. Dots indicate estimations of the individual participants' parameters. B-D) Interactions of experimentally controlled factors. E-H) Between subject factors, note the different scale for F-H and that these are slopes (effects per unit) and not categorical effects.

5.3.2   Features of the assessment

We turn now to the effects of the assessment methodology on the behavior, shown in Fig 5.3. These effects are to be interpreted as interactions on, or modifiers of the global preferences presented in the previous paragraph, and tell us how the behavior changes if we change the respective aspect of the assessment.

Level of abstraction: Going from a text-based to a naturalistic presentation showed no conclusive effect on the omission bias (study 1: -1.3, CI_{95}: $-2.9, 0.3$, BF_{H0}: 1.1; study 2: -0.6, CI_{95}: $-1.5, 0.3$, BF_{H0}: 2.9), but strong evidence for higher valuation of elderly people in study 2 (study 1: 1.8, CI_{95}: $-0.3, 4.0$, BF_{H0}: 0.7; study 2: 4.6, CI_{95}: $3.0, 6.3$,

BF_{H0}: 0.0). To determine whether the inclusion of the level of abstraction as a predictor improves our model fit, we used the brms package [17, 18] to calculate the difference in Watanabe–Akaike Information Criterion (WAIC) between the full model, and an identical model without any abstraction-related fixed effects (using the same random structure). Study 1 shows slight evidence against the more complex model (Table 5.1). Since the standard error of the difference is small compared to the magnitude of the difference, the simpler model without level of abstraction as a factor is the superior model here. Study 2, on the other hand, shows slight evidence in favor of the more complex model (Table 5.1). However, the magnitude of the difference is smaller than in the first study, and the standard error of the difference is larger, giving more credibility to the findings from study 1. Moreover, the observed influence of level of abstraction as a whole is likely driven by the interaction between level of abstraction and elderly bias. We therefore conclude that the level of abstraction has no significant influence on the participants' decisions outside of the valuation of the elderly.

Table 5.1: **WAIC model comparison: Level of abstraction**

	study 1		study 2	
	WAIC	SE	WAIC	SE
with Abstraction	598.4	35.5	1141.85	50.1
without Abstraction	570.2	35.6	1118.3	51.2
with - without	28.2	5.77	-23.6	10.5

Model with fixed effects of level of abstraction against model without them, for both studies individually.

Modality: Contrasting an immersive VR environment to a desktop setting with static scenes and unlimited response times did not yield any conclusive effects, except strong evidence that the lane bias is independent of the modality (0.2, CI_{95}: $-0.5, 0.8$, BF_{H0}: 8.3). We can further exclude large effects of modality on the gender bias (-0.7, CI_{95}: $-2.1, 0.7$, BF_{H0}: 2.7). For the young and elderly interaction we observe a bimodal distribution in our data (see Fig 5.3), which leads to large credibility intervals of the effect for the interaction of young (1.0, CI_{95}: $-2.3, 4.4$, BF_{H0}: 1.5) and elderly (2.0, CI_{95}: $-0.9, 5.0$, BF_{H0}: 0.8).

Time pressure: High time pressure, on the other hand, did have a considerable impact on the decision patterns, as it led to systematic decreases of the age bias (young: -3.0, CI_{95}: $-4.8, -1.4$, BF_{H0}: 0.0; elderly: 2.5, CI_{95}: $0.8, 4.3$, BF_{H0}: 0.1) as well as a trend towards lower gender bias (-1.0, CI_{95}: $-2.0, 0.1$, BF_{H0}: 1.1), but with an inconclusive Bayes factor. Overall, this result is in line with more randomness in the participants' answers, or in other words, an increased error rate. However, it was not caused by a simple failure to elicit a response in time. Running out of time to enter a response would lead to fewer lane changes overall and cause an increased omission bias, both of which we did not observe (omission bias: -0.2, CI_{95}: $-1.1, 0.7$, BF_{H0}: 5.9; lane changes see S3 Appendix: Trials and error rates). However, we cannot discern whether, or to which extent, this systematic effect is the result of errors, such as pressing the wrong button or misidentifying an obstacle's gender or age under time pressure, or the result of interrupting the cognitive process of evaluating ethical aspects of the situation. Mind that these effects don't seem to carry over to the slower VR conditions with 4.0-4.4s response time windows. The absence of systematic effects between time-constrained and unconstrained modalities, and the fact that 72% (text-based) to 82% (naturalistic) of all responses in the unconstrained settings were made within the response time windows of the time-constrained settings (see S3 Appendix: Trials and error rates), suggest that response time windows of about 4.0s are not restricting the validity of immersive VR-based assessment. We conclude that aside from very high time pressure, the observed decisions are remarkably consistent across different approaches to the experimental assessment.

5.3.3 Individual features of the participants

This third set of features refers to individual differences between the participants, namely their age and gender, as well as video gaming experience and susceptibility to social desirability. The influence of these features on the behavior is again modeled as interactions with the global preferences, to be interpreted as deltas on them.

Gender: Female participants showed a tendency to value elderly people higher than male participants did (-1.8, CI_{95}: $-4.2, 0.6$, BF_{H0}: 0.8) but the Bayes factor is inconclusive. Interestingly, we found (weak) evidence for no effect between male and female participants with respect to gender bias (-0.4, CI_{95}: $-1.9, 1.1$, BF_{H0}: 3.6), supporting the notion that a pro-female gender bias is generally agreed upon in the sample population.

Age: We modeled the influence of age using the continuous age predictor with several interactions, and the resulting parameter estimate is to be interpreted as change in odds per year. We found evidence against any effect of participants age (all BF_{H0}:> 12).

Video game experience: The amount of video game experience (measured in game playing hours per week) had virtually no influence on any of the parameters, rejecting the hypothesis that frequent players could have a different conception of the stakes involved in these scenarios (all $BF_{H0}:> 17$).

Social desirability: Higher scores in the SDS-17 characterize an increased tendency to respond in line with social norms instead of one's own true beliefs. We found evidence against any moderate or large effects of SDS-17 scores (all $BF_{H0}:> 7$), but the credible intervals indicated trends towards small effects in omission bias (0.18, CI_{95}: $-0.05, 0.42$, BF_{H0}: 7.3) and pro-female gender bias (0.16, CI_{95}: $-0.08, 0.40$, BF_{H0}: 10.4). People with a stronger tendency to be influenced by social norms may thus prefer not to take action, in order not to increase their perceived own guilt, and a higher valuation of females would arguably be in line with social norms in modern western societies. However, our analysis makes anything but small effects in terms of omission and gender bias unlikely.

5.4   Discussion

We conducted two studies to show whether different assessment methods change the ethical decisions of participants in road traffic dilemma situations. Our main finding is that by and large this seems not the be the case, and VR and text-based assessments appear to measure the same underlying construct, with only minor shifts in behavior between the different methods. By extension, this supports the suitability of the methodology and comparability of the results obtained in previous studies [109, 11, 39, 43, 83, 5].

Personal involvement has previously been linked to "reduce[d] sensitivity to moral norms and [an] increase[d] general preference for inaction" [44]. In an exploratory post-hoc questionnaire in study 2, participants reported they could more easily put themselves in the presented situation in the naturalistic trials, than in the text-based ones (naturalistic: 5.1/7; text-based: 4.2/7). This would lead to the assumption that the omission bias should increase in naturalistic settings. On the other hand, taking the driver's perspective in a naturalistic setting may cause participants to perceive both options as actively causing harm, thus taking away the supposed moral superiority of inaction. However, while the sign of the posterior means indicated a slight reduction in omission bias in both studies, our results were inconclusive with regard to this question. The most pronounced difference between the naturalistic and text-based representations was a higher valuation of the elderly in naturalistic settings. It is possible that the abstract, text-based representation of "elderly" strips the human person of all other attributes, making age more salient than it would otherwise be. This effect of age could, therefore, be an artifact of the abstract presentation. Some participants, however, reported needing longer to distinguish between elderly and adults in the naturalistic environment than in the text-based setting. The observed effect could, thus, also be an effect of the elderly's particular visual representation in the virtual environment. Participants may have perceived them closer in age to adults, resulting in a higher valuation. If a naturalistic environment led to a generally reduced influence of the victim's age on the decisions, we would expect the value of young people to decrease in this setting, which it did not.

The level of abstraction in the presentation was experimentally detached from its modality, i.e., whether the decisions were made in a VR environment under time-constraints, or in a regular desktop setting without any time-constraints. While we can exclude large effects of modality on gender bias, our findings on the omission and age biases were inconclusive.

Severe time pressure, i.e., response time windows of 1.2 and 1.6 seconds, respectively, caused a systematic decrease of gender and age biases, consistent with the notion of erroneous identification of the potential victims, or interruption of the cognitive evaluation of the situation. However, no such effects were found between conditions of 4.0 second and unlimited response windows, suggesting that response windows of about 4 seconds are not long enough to have a considerable impact on the participant's decisions in the presented scenarios. Time constraints have previously been found to influence moral judgments in some cases [108], but not in others [119, 7]. When found, such differences are typically viewed as evidence in favor of the dual process theory, which links fast and intuitive cognitive processes to deontological reasoning, and slower cognitive processing to utilitarian reasoning [57]. If we interpret deliberate inaction as a deontological choice, in which the norm of not actively causing harm trumps a higher perceived valuation of the obstacle in the given lane, we could construe the lack of a difference in omission bias between the fast and slow condition as evidence against the dual process theory. However, such an interpretation is difficult for two reasons. (1) Equating inaction with deontological judgment is problematic, since the two may generally represent independent factors in moral decision making [44]. (2) Even if we allow this equation, it is not clear whether the omission of a lane-change would be perceived by the participants as refraining from active interference, since the active operation of the car may negate this notion.

What does this mean for future studies? The assessment of more complex and dynamic traffic situations could benefit from the use of VR, in cases where the experimental situations become difficult to fully or precisely explain in text or still images. At the same time, VR assessment is rather costly, requiring specialized hardware and substantial development time to create the applications. The assessment itself is also cumbersome in comparison to abstract, possibly browser-based surveys. The relatively low cost and the ease of reaching large numbers of participants, as exemplified in the Moral Machine [5], make simple and abstract presentations of the scenario the economically preferred choice.

Outside of the examination of different approaches to the assessment, our findings largely support earlier studies [5, 39, 11, 43, 83, 109], forming a coherent picture of the ethical principles in our society, as they apply to traffic dilemmas. Unsurprisingly, animals are generally valued far inferior to humans. When having to decide between multiple potential human victims, the utilitarian principle of minimizing the overall harm appears to outweigh most other factors, while a tendency towards inaction only plays a minor role in the decisions. Beyond these, the potential victim's age appears to be the most decisive factor, followed by their gender. A quantitative comparison between the different studies is, unfortunately, difficult to obtain, due to different experimental setups and modeling approaches. Notably, the observed differentiation by age, gender, and other personal factors stands in contrast to the ethical rules outlined in the report of the ethics commission of the German Federal Ministry of Transport and Digital Infrastructure [88]. In their report, sacrificing an innocent person for the greater good is viewed as strictly unacceptable, and even basing a decision on the number of lives saved between already involved parties is met with severe ambivalence. The findings in this and previous studies thus highlight severe points of contention that need to be addressed by manufacturers and legislators, since they may affect public acceptance of automated driving technology.

With respect to personal factors influencing the decisions, we found no difference in the pro-female gender bias between male and female participants, attesting a high consensus on this aspect. This finding is at odds with the findings of the Moral Machine, where female participants were found to have a much stronger pro-female gender bias than males [5]. This difference may be attributed to the non-representative sample on our side, consisting mostly of undergraduate university students.

We further found females to value elderly people higher than males did, thus having a smaller overall age bias, while the Moral Machine found female respondents to have a slightly larger age bias than males. A possible explanation for this discrepancy lies in the use of a single scale for age bias in the Moral Machine, which could be masking the differential valuation of individual age groups we report. Beyond this, we found no notable effects of the age of the respondents or their experience with video games. A factor that has not been accounted for in previous traffic dilemma studies is that of social desirability. We found marginal evidence for higher scores on the Social Desirability Scale (SDS-17) predicting larger omission and gender biases, creating a leverage point for future studies in this field.

Outlook

Future studies could address a factor that was only partially discussed in earlier work: When participating in road traffic, be it as a pedestrian, cyclist, or car driver, we consent to a certain level of risk depending on our actions. For instance, common sense dictates that the safety of pedestrians on the sidewalk takes precedence over pedestrians stepping into the street or even jaywalking. This introduces the aspect of fairness to the question, which trajectory to select or whom to put at risk in a critical situation. Since the written law typically does not provide a nuanced conception of consent and fairness, the topic is a prime candidate for empirical assessments. Aside from considerations of fairness, the individual options in dilemma situations will often have different levels of expected collision severity, or expected speeds at impact. This might substantially influence one's moral assessment of a given situation, but hasn't been systematically addressed in traffic dilemma studies so far. We believe that these aspects could provide valuable insight into our moral intuitions as they relate to road traffic and possible solutions for ethical decision making in self-driving cars.

5.5   Conclusion

In conclusion, the present work establishes the general comparability of trolley-like traffic dilemma studies using various methods of assessment. It further substantiates previous findings on global preferences guiding our decisions in these scenarios, helping to inform regulation, communication, and possibly implementation of ethically sound decision-making systems in self-driving cars.

	study 1	study 2
participants recruited	88	107
excluded from analysis	3	14
remaining	85	93
females	43 (50.6%)	58 (62.4%)
males	42 (49.4%)	35 (37.6%)
age	23.0 (SD 3.6; 18-40)	21.3 (SD 3.8; 18-47)
using German language	85 (100%)	88 (94.6%)
using English language	0 (0%)	5 (5.4%)
VR setting	43 (50.6%)	93 (100%)
desktop setting	42 (49.4%)	-
selection criteria		aged 18+
		no prior participation in similar studies
		no traumatic traffic-related experiences
		no epilepsy or psychotic disorders
further information	mostly university students (cognitive sc., psychology)	

Table 5.2: Sample descriptions for both studies.

5.6   Supporting information

S1 Table: Sample overview.

S1 Appendix: Conditions and controls.   For the naturalistic VR setting, we used a custom 3D environment portraying a suburban residential area. The participants were virtually placed in the driver's seat of a compact car, heading down the road at 36 kph ($\approx$22 mph), and wore headphones playing engine, road and wind noises, as well as a tire rubbing sound during lane changes. The car's starting lane was pseudo-randomly assigned. A wall of dense fog was placed at a fixed distance of about 50m to the car, such that after the obstacles appeared from the fog, participants had exactly 4.0s to decide which lane they would choose, and no braking or other ways of preventing a collision were possible. Inputs were made using the keyboard's arrow keys, and the time remaining to make a decision was indicated acoustically by a series of four beeps, the last one of which had a higher pitch and marked the end of the decision time frame. At this point, control over the car was taken from the participants, and the car went on for another 10 meters in order to complete any ongoing lane changes. The car stopped abruptly right before impact with the obstacles, at which point all sound was muted and the screen faded to black, marking the end of the trial.

In the text-based VR setting, the immersive 3D environment was replaced by an abstract visualization of the same scenario. The chosen lane was indicated by white lane markings on an otherwise uniformly gray background, and the obstacles were indicated as white text. The obstacle currently at risk was positioned centrally above the lane markings and was additionally marked with a small arrow, while the other obstacle was placed to its left or right. Lane changes were indicated by shifting the arrow to the other obstacle, and moving the obstacle texts such that the new obstacle at risk would again end up centered above the road markings. The decision time in this condition was set to 4.4s. The additional 0.4s over the naturalistic condition served to make up for the longer time required to read and comprehend the situation in text-based settings, with the aim of aligning the true decision time and perceived time pressure between the conditions. The precise value was estimated and agreed upon in a pre-study assessment with four participants. The same auditory cue as in the naturalistic condition was used to indicate how much time was left to make a decision. Additionally, the lane markings were progressively covered up over the course of the 4.4s, resembling a declining progress bar.

In the desktop conditions, participants were presented with a questionnaire that either used a still 3D rendering (naturalistic), or a text-based description of the situation. In both cases, participants were asked the same question: "Given only the choice between the left-hand and the right-hand lane, in which of the two would you choose to drive?" In contrast to the VR conditions, the desktop conditions had no default answer, but showed two open circles to select from. The selection was done via the left and right arrow keys and could be changed back and forth until it was confirmed by pressing the space bar. Importantly, there was no default option, and no time limits were imposed in the desktop conditions. While this design decision did not allow us to fully disentangle the influence of the modality from that of time pressure, it did allow us to examine the response times in the absence of any time pressure, allowing us to gage the amount of time pressure imposed with a four second time limit. Moreover, an unlimited time frame puts these conditions more in line with browser-based questionnaires used in the literature.

The slow conditions in the second study were identical to those in the first study, while the fast conditions had the reaction time window reduced to 1.2 and 1.6 seconds for the naturalistic and text-based conditions, respectively, maintaining the 0.4 seconds of extra time for the text-based settings. In the naturalistic condition, this was achieved by increasing the car's velocity twofold and shortening the fog distance accordingly. In the text-based condition, all visual indicators were sped up to match the shortened time frame. The acoustic indicators (a series of four beeps) were also adjusted accordingly in both conditions.

S2 Appendix: Timelines and hardware. **Study 1.** Prior to the experiment, the subjects were informed that they had the option to stop the experiment at any time, that their decisions would not be monitored in any way and would be saved anonymously, and signed a consent form. Participants in the desktop screen setting sat in front of a 24" PC screen, whereas in the VR setting, participants wore an HMD (HTC Vive), as well as a pair of headphones (Sennheiser 380 Pro or Bose QuietComfort 25). Regular PC keyboards were used for all user inputs, and participants wore electrodes of an EDA measuring device on two fingers of the left hand. The participants were asked to place their right hand's fingers on the arrow keys and their left hand's fingers on the space bar, since no other keys had to be used in the experiment. In the application, the experimental trials were preceded by a form asking for the participant's age and gender, and a text explaining the controls and introducing the upcoming experimental condition. To give participants in the VR setting a chance to familiarize themselves with the virtual environments, test trials were conducted before each of the two main experimental conditions (naturalistic and text-based), where a number of traffic cones had to be avoided. When a traffic cone was hit, the training trial was repeated until all cones were cleared in a single run. The total time spent in the experiment was about 12 to 15 minutes for participants in the VR setting, and about 5 minutes for those in the desktop screen setting. After completion of all trials, the participants were interviewed about the immersion in the different scenarios and the correct apprehension of the presented task, using standardized questions.

Study 2. The procedure for study 2 was mostly identical to the VR setting in study 1. The total time participants spent in the application was also between 12 and 15 minutes. An exploratory post-hoc questionnaire with questions about the apprehension of the presented task had to be filled in within the application after the experimental trials. Finally, all participants filled in a paper-based form of the SDS-17 questionnaire assessing social desirability.

S3 Appendix: Trials and error rates. **Study 1.** A total of 3400 trials were analyzed in the first study, i.e., between 840 and 860 in each of the four conditions. Of these, 1136 trials (274 to 295 per condition) were direct juxtapositions of humans. In order to estimate the frequency of erroneous trials in our data, we asked participants in a post-hoc questionnaire to recall the number of errors they had made during their trials (e.g., accidentally hitting the wrong button). A total of 30 errors were reported for 3400 trials (0.9%). Furthermore, we took the rate of hitting an obstacle when the other lane was empty as a second measure,

	trials	switches	trials with switches	h. vs h. trials	error rate	RT mean (SD)	% RT under VR limit
VR naturalistic	860	945	503 (58.5%)	295	0.54%	1.42s (1.02s)	-
VR text-based	860	586	450 (52.3%)	278	0%	1.65s (0.79s)	-
Desktop natural.	840	-	-	274	0%	2.60s (3.08s)	82.4%
Desktop text-b.	840	-	-	289	0.64%	0.64s (3.85s)	72.2%
total	3400	-	-	1136	0.29%	-	-

Table 5.3: Study 1. Left to right: Total number of trials, total number of lane switches, number of trials in which lane switches were performed, number of trials with humans in either lane, error rate estimate (rate of "saving" the empty lane), response time means (VR: only taking trials with lane switches into account), percentage of responses within the time limit of the corresponding VR conditions (naturalistic: 4.0s, text-based: 4.4s).

	trials	switches	trials with switches	h. vs h. trials	error rate	RT mean (SD)
VR naturalistic slow	651	645	363 (55.8%)	465	0%	1.62s (1.11s)
VR text-based slow	651	363	328 (50.4%)	466	0%	1.55s (0.73s)
VR naturalistic fast	651	526	371 (57.0%)	468	1.11%	0.61s (0.29s)
VR text-based fast	651	396	348 (53.5%)	474	1.70%	1.02s (0.25s)
total	2604	1930	1410 (54.1%)	1873	0.55%	-

Table 5.4: Study 2. Left to right: Total number of trials, total number of lane switches, number of trials in which lane switches were performed, number of trials with humans in either lane, error rate estimate (rate of "saving" the empty lane), response time means (only taking trials with lane switches into account).

which was observed in 2 out of 688 trails (0.3%). We take from these figures that errors in the response are unlikely to have a strong impact on the results. Moreover, we found animals to be saved over humans in only 17 out of 1354 trials (1.3%). Since this rate is very close to the error rate estimators, it seems plausible that species is an almost trivial factor within the participants' demographic. We, therefore, decided to focus on trials with humans in either lane for the behavioral analysis of study 1.

Study 2. In the second study, a total of 2604 trials (651 per condition) was recorded, 1873 of which were juxtapositions of humans. As an estimation of the error rate, we again used the rate of trials in which a human was hit despite the other lane being empty. This never occurred in the slow conditions, but happened in 1.7% and 1.1% of trials in the fast text-based and fast naturalistic condition, respectively. Note that since the task of correctly identifying age and gender of the obstacles is arguably harder than identifying the mere existence of an obstacle, misinterpretations of the scene could have caused the true error rate to be higher than the estimator under severe time constraints.

S4 Appendix: Bayesian logistic regression model specifications.

Study 1. The dependent variable for all models was whether subjects choose the left lane (1) or the right (0). For the first study, we used as predictors the intercept (*1*; coding for a lane-bias towards the left lane), the lane the participants were in at the moment of the obstacles' visibility onset (*omission.bias*; coding for a bias to stay in the lane), the configuration of obstacles in the two lanes, as well as the *modality* and *abstraction* level of the respective experimental condition and their interactions. Note that there was no default choice indicated in the desktop conditions, and thus no omission bias or effect of the modality on an omission bias can be calculated. The configuration of obstacles in the two lanes was analyzed similar to Bradley-Terry-Luce models [14] and was coded as follows: *gender.bias* is 0 if both obstacles have the same gender, .5 if the left obstacle is male and the right obstacle is female, and −.5 vice versa. *young.bias* is 0 if both obstacles are children, .5 if the right obstacle is a child but the left obstacle is not, and −.5 vice versa. *elderly.bias* is the corresponding equivalent for the elderly. The adult obstacles weren't explicitly modeled, as it would result in an over-specification of the model. The model can thus be understood as fixing the model parameters for adults at 0 and modeling the other two age groups relative to this. *Modality* (VR/ desktop) and *abstraction* (text-based/ naturalistic) used effect coding, i.e., the previously specified effects do not represent the effect at one specific condition, but at the average over the four conditions. The predictions for any particular condition are obtained by adding or subtracting one half of the respective modality and abstraction parameters. All models used weakly regularizing priors. For parameter means, we used Normal($\mu = 0, \sigma = 3$), for the variance of the distribution of parameters by subject we used Cauchy($x_0 = 0, \gamma = 1$), and for co-variance matrices we used LKJ($\eta = 2$). For all analysis we made use of the BRMs package [17, 18] using the NUTS Hamiltonian Monte Carlo algorithm [65, 20]. We used 8000 samples for each chain including 2000 warm up samples with 4 independent MCMC chains. We report 95% credibility intervals of the posterior and the mean posterior value. Bayes factors were calculated using the Savage-Dickey density ratio method as implemented in the brms package. We interpret Bayes factors between 1/3 and 3 as mostly inconclusive, Bayes factors above 10 as strong evidence. The model for study 1 was specified as follows:

$$
\begin{aligned}
\text{choice.left}_{s1} \sim\ & 0 + \text{Intercept} \\
& + (\text{omission.bias} + \text{gender.bias} + \text{young.bias} + \text{elderly.bias}) * \text{modality} * \text{abstraction} \\
& + (0 + \text{Intercept} + (\text{omission.bias} + \text{gender.bias} + \text{young.bias} + \text{elderly.bias}) * \text{modality} * \text{abstraction} | \text{subj.idx})
\end{aligned}
$$

Study 2. The model used in the second study is based on that of the first study, with the following changes: *Speed* replaces *modality* and was coded in the same $-0.5/0.5$ scheme. In addition to the features of the portrayed situation and the features of the experimental condition, we included features of the individual subjects in the model: *subj.gender* is the gender of the subject in the trial (using effect coding, i.e., ± 0.5), *subj.age* is their age (variable centered, reported effects are effects per year of age), *subj.gamehrs* is their reported weekly average of hours spent playing video games (variable centered, reported effects are effects per one hour of weekly playtime), and *subj.sds17* is the score of their SDS-17 questionnaire (variable centered, reported effects are effects per one score in the test). The model was specified as follows:

```
choice.left_S2 ~ 0 + Intercept
        +(omission.bias + gender.bias + young.bias + elderly.bias) * speed * abstraction
        +(0 + Intercept + (omission.bias + gender.bias + young.bias + elderly.bias) * speed * abstraction|subj.idx)
        +(omission.bias + gender.bias + young.bias + elderly.bias) * subj.gender
        +(omission.bias + gender.bias + young.bias + elderly.bias) * subj.age
        +(omission.bias + gender.bias + young.bias + elderly.bias) * subj.gamehrs
        +(omission.bias + gender.bias + young.bias + elderly.bias) * subj.sds17
```

S5 Appendix: Regression coefficients. Study 1 . All $\hat{R}$ values were below 1.1 indicating convergence. Due to the high number of predictors only a random selection of predictors was visually inspected for convergence. We noticed that in study one, the second level standard deviation marginal density for the effect "elderly.bias" was bimodal (but not the fixed c oefficient). No ne of th e co rrelations on th e population level showed indication for an effect. It is very likely that our model was too complex to estimate such correlations reliably given our sample size and the lkj-prior takes over. Table 5.5 reports the fixed effects conditional on the population level. The main effects of the coefficients are all very large, indicating a binary decision by most subjects.

Study 2. The results of study two, shown in Table 5.6 are very similar. No bimodality was found in the marginal posterior densities.

Chapter 6

Towards a Framework for Ethical Decision Making in Automated Vehicles

Contributions

PsyArXiv: Leon René Sütfeld, Peter König, and Gordon Pipa 2019
    LRS: Conceptualization, Investigation, Project administration, Visualization, Writing – original draft; PK: Supervision, Writing – review & editing; GP: Supervision, Writing – review & editing

Layman's Summary

This paper concludes the topic of ethical decision making in this book by exploring the status quo of the debate on algorithmic decision making in automated vehicles, and raises a number of open issues in the way of a framework for regulation and implementation of ethical decision making algorithms in AVs.

We first motivate the discussion on *how* to perform ethical evaluations in automated vehicles by laying out *why* constant ethical evaluations of the scene around the vehicle are going to be necessary. In the core part of the paper we then perform a review of existing ethical guidelines and demands, and most prominently the report of the ethics commission for automated and connected driving, convened by the German Federal Ministry of Transport [88]. We analyse key shortcomings of these demands, and explore possible solutions. Among the central issues raised here are the necessity of making decisions based on probabilistic data, i.e., dealing with uncertainty about the outcome of a given situation at decision time, and the incompatibility of strictly categorical rule systems with robust, reasonable and transparent decision making. Further, we touch on the possibility of comparing all ethically relevant factors of a situation on a single ethical scale, and we discuss metrics to rate a person's involvement in a given traffic situation and their responsibility for the creation of risk. We conclude the central part of the paper by deriving 10 central findings from this discussion, and point out which issues, to our belief, are most consequential in the current debate. Finally, we give an outlook on possible technical solutions that could be derived from a coherent framework for ethical decision making in AVs – ethical evaluations of continuous space as opposed to the analysis of a low number of individual route choices, and based on this, the optimization of trajectories in regular driving.

6.1   Introduction

Driven by fast progress in artificial intelligence, automated driving technology has in recent years become a key area of research and development in the automobile industry. However, getting to the point where automated vehicles (AVs) can be trusted to drive on public roads without human supervision isn't just a question of technology. It brings with it a number of ethical and regulatory challenges that need to be addressed in order to facilitate a smooth launch of the technology, avoid costly legal disputes and public backlashes, and ultimately increase road safety [88, 12].

Many of the open ethical questions concerning automated driving revolve around route choices, handling of user data, or biases in AI-driven perception systems [66, 125]. Most prominently debated, and at the heart of this paper, are decisions that AVs may have to make in response to dangerous situations, or imminent collisions. Take, for example, a situation where a driver turns left at an intersection and overlooks an oncoming AV. The AV can only avoid a collision by swerving into oncoming traffic itself, and has to make a decision without knowing exactly how the events will play out. Staying in its lane may result with near certainty in a severe collision, while swerving into oncoming traffic could lead to a collision with another car, or that car may be able to avoid the collision with an emergency maneuver. This example alone raises a number of questions: How should probabilistic predictions be handled? Can the AV be at all allowed to swerve into the opposite lane, committing a traffic violation? And to what extent is it permissible to put otherwise uninvolved parties at risk? A number of trolley-like studies have shed a light on human behavior in such situations, which create a frame of reference as to what we deem morally right behavior [5, 109, 112, 11, 39]. But many of the biases we show as humans in our personal judgment of specific situations are at odds with the moral foundations of our law and society. So while these studies can inform the debate, the observed behavior can't simply be copied over to AVs [110, 88].

Another aspect of ethical behavior in road traffic are trade-offs between utility and safety. They include the adaptation of the driving speed to the immediate environment of the car, determinations of when it is safe enough to overtake a cyclist or another car, and even the precise positioning of an AV within its lane. Arguably, both common and critical situations would benefit from, or even require, an ad-hoc ethical evaluation of a given situation, and a set of behavioral principles that govern the AV's actions based on this evaluation. What these evaluations should look like, and what behavioral principles should govern the self-driving vehicles' behavior, however, is an unsolved problem to date.

6.2 The Case for Ethical Decision Making in AVs

Before we discuss the central issues for the design of ethical decision making systems in AVs, we will motivate this debate by laying out when and why such systems become a necessity.

6.2.1 Levels of Automation

At what point ethical decision making becomes relevant for AVs can be deduced from the Society of Automotive Engineers's (SAE) categorization in six levels of automation, ranging from no automation (level 0) to fully automated (level 5) [102]. On the first three levels, 0-2, the task of monitoring the environment and responding to events whenever necessary lies with the human driver. Level 1 and 2 systems are consequently categorized as driver support features. While these can include limited response to critical situations, i.e., emergency braking, they are not designed to be relied on, and the driver is supposed to intervene whenever necessary to guarantee safe operation of the vehicle. The responsibility to make ethical decisions, therefore, still lies with the driver. From level 3 on, the task of monitoring the environment, as well as performing collision avoidance, lies with the system. Such systems are classified as Automated Driving Systems (ADS), and no longer require the driver, or user, to be attentive to the road. While on level 3 the driver needs to be fallback-ready, "sufficient time for a typical person to respond appropriately"[102] has to be provided when issuing a request to intervene. Since this will take at least 5 seconds for an average driver [37], it is not an option for any immediately dangerous situation. The task of any required ethical assessments and decision making is thus passed on to the ADS for levels 3 to 5.

The case can be made that a level 3 or level 4 system would not necessitate ad-hoc ethical assessments, if automated driving is restricted to specific driving domains in which critical situations that require thorough ethical evaluation are exceedingly rare. This is a valid point, provided sufficient statistical support that the driving domains present highly predictable low risk scenarios. For most country roads, and in particular for all suburban and urban areas, however, such exceptions seem out of the question. Finally, level 5 is characterized by an ability

to operate the car under all circumstances and in all domains in which
a human driver could operate it, and consequently can not be exempt
based on the available driving domains.

6.2.2   Need for Ethical Decision Making in AVs

A common argument against the need for ethical decision making
systems in AVs is that these will rarely cause any accidents, with hopes
of up to a 1000-fold reduction in accident frequency [103]. To see if this
claim has merit, we can adduce the safety levels of both AV prototypes
and currently available level-2 systems as a frame of reference.

Favarò et al. [40] report on accident statistics by the California De-
partment of Motor Vehicles, showing that AV prototypes between 2014
and 2017 caused accidents at a similar rate as human drivers. Notably,
unexpected braking by the prototypes lead to a very high number of
rear-end collisions, driving up the total rate of accident involvement
10-fold. With respect to safety in commercial releases, Tesla set a bench-
mark with their late 2015 release of Autopilot, a level-2 ADAS restricted
to highway use. Unfortunately, reliable data on the standard of safety
of the Tesla Autopilot system is difficult to obtain. Statistics provided by
the US Insurance Institute for Highway Safety were analyzed by Mar-
shall [92] and Thomas [116], who estimate that Teslas equipped with
Autopilot show a slight reduction in accident rates of up to 35% (~ 1.5-
fold). However, the data did not differentiate between manual driving
and Autopilot mode, and the numbers may be confounded with other
safety-improvements.

These data suggest that, at least for the first generation of cars
equipped with level 3 systems or higher, safety levels will be far from
the goal of a 1000-fold reduction compared to human drivers. Moreover, there is an incentive not wait for the technology to become orders
of magnitude safer than human drivers before it is introduced to the
market: The RAND Corporation [100] estimates that even an introduction with safety levels on par with that of humans could save thousands
of lives in the long run, due to a faster advancement of the technology
when large amounts of data are available sooner. We thus find it reasonable to assume that AVs' safety levels will be comparable to, or only
moderately better than human drivers, at the time of their introduction.

Importantly, even a large safety-improvement in AVs would not keep
them from being involved in accidents caused by other road users. Without an ad-hoc evaluation of the risk factors involved in a critical situation, any AV would have to be programmed to strictly stay in its lane
and to use emergency braking as the only means to react to a dangerous

situation. A balancing of multiple sources of risk, i.e., risk management [52], and an ethical evaluation of the involved factors will often be necessary to sensibly avoid or mitigate collisions, regardless of who is at fault for the situation. We thus argue in line with Goodall [52] that ethical assessments and appropriate risk management algorithms should be regarded a necessity for AVs, and qualify to be mandated by legislation.

6.2.3 Safety Potential of Ethical Decision Making Systems

This leads us to the question how often AVs can be expected to end up in a situation where the described mechanisms would noticeably improve the outcome of a critical situation. The German Federal Office of Statistics annually publishes detailed accounts of crash statistics for Germany [16]. In 2017, a total of 2.6 million accident reports were filed, with roughly 300.000 (12%) of them entailing at least minor injuries to one of the involved persons. The latter are assessed and analyzed in more detail in the report. Based on these numbers, we make a projection with respect to safety-benefits of ethical decision making systems for two different future scenarios. The first scenario is set in the medium term future and assumes 5% of all kilometers to be driven by AVs, and AV accident rates 5 times lower than humans'[1]. The second scenario is set further in the future, with 50% of all kilometers driven by AVs, accident rates among human drivers cut in half, and AV accident rates 1000 times lower than those of human drivers in 2017. Across all kinds of accidents, we assumed that in any accident with AV involvement there is a 10% chance that an ethical assessment of the situation would allow the AV to change its trajectory to achieve a "better" outcome of the situation. This could mean a less severe collision with reduced bodily harm to those involved, or a fairer outcome from an ethical point of view. We further assume that in a quarter of these cases, a collision can be prevented altogether.

The results are summarized in table 6.1, and all estimations are laid out in detail in appendix 6.8. Most importantly, the mid-term future scenario already yields an annual 12, 920 accidents with human injuries in which AVs are involved. Among these are an estimated 1, 163 cases in which the outcome could at least be improved, and 293 accidents that could be avoided altogether. In the far future scenario, these numbers increase by a factor of 2.7. Keeping in mind that these are estimates for Germany alone, this outlines an enormous potential for improved safety and more ethical, i.e., "fairer" behavior, when AVs are equipped

---

[1]These numbers were contrived exemplarily to provide conservative estimates for the expected number of accidents with AV involvement.

with systems performing real time ethical evaluations as part of their decision making process.

Scenario	Total	AV Involved	Caused by AV	Outcome Improved	Prevented
2017	302,656	-	-	-	-
near future	291,247	12,920	2,756	1,163	293
far future	85,473	34,009	143	3,163	791

Table 6.1: Annual number of accidents with personal injuries (total, with AV involvement, and caused by an AV) in the estimated near and far future scenarios (5% and 50% of all kilometers driven by AVs, respectively.) *Outcome Improved* and *Prevented* refer to the assumed safety potential of ethical decision making systems. Table constitutes summary of the tables found in appendix 6.8.

6.3 Towards a Framework for Ethical Decision Making in AVs

Despite the need for ethical decision making systems in AVs, there is no suitable framework to date that could serve as the basis for a practical solution. We will, therefore, review existing guidelines and demands with regard to their theoretical and practical feasibility, and based on this compile a list of 10 requirements and open questions for the implementation and regulation of ethics in AVs.

A precedent with respect to fundamental moral rules is set by the report of the ethics commission for automated and connected driving, convened by the German Federal Ministry of Transport [88]. As it is the first document of its kind issued by a federal ministry, it stands as a point of orientation for other lawmakers. Although their views aren't legally binding, we are not aware of any contending ethical statements from official sides at the time of writing, and, therefore, treat these demands as a starting point in this discussion.

6.3.1 Need for Explicit Regulation

Existing regulations touching upon ethical aspects of certain traffic situations, which were conceived to regulate the traffic behavior of human drivers, cannot be applied in the same way to AVs. As a case in point, the German ethics commission states that "[a]lthough a human driver would act illegally if they killed a person in an emergency to save one or more other people, he would not necessarily be held guilty. Such judgments, employed in retrospect for special circumstances cannot be converted without further ado into abstract-general ex-ante judgments and thus also not into appropriate programming"[88]. In other words, while a human driver can be forgiven after making a questionable decision in an emergency situation, the same reasoning cannot be applied to AVs. An AV makes trajectory decisions based on information about its own state and its immediate environment, collected and combined into a world map by its perception module [60]. It applies a pre-defined set of behavioral principles, specifying the car's actions as a function of the perceived state, the desired route, and other factors. These behavioral principles also define how the AV would react in an emergency situation. Since they are pre-defined in the programming, any morally reprehensible decision would have to be construed as premeditated. Moreover, since these behavioral principles are known at the time of the AV's certification, it would be difficult to argue any wrongdoing on the side of the manufacturers after the fact for behavior that arises from these principles. By virtue of the precautionary principle, regulatory bodies thus have an obligation to verify that the implemented behavioral principles are morally acceptable. Consequently, in-place regulations conceived for human drivers do not suffice and the regulations for automated driving need to specify explicitly what constitutes morally acceptable behavior[2] (§1).

6.3.2 Transparency

An important requirement of the ethics commission's report is for the public to be adequately informed about the decision making systems [88] (§2). Birnbacher and Birnbacher [12] take the same line, arguing that ethical decisions should under no circumstances be made by black box algorithms, and that the implemented moral norms should ideally be understood and shared by everybody. Although the demand for transparency appears natural, its implications should not be underestimated. The precise decision making process of human drivers in critical situations is concealed by severe time constraints, and their ethical assessment of the situation is usually not precisely explicable, nor put under scrutiny. As a consequence, human drivers can proclaim to generally act according to specific ethical principles (e.g., non-discrimination with respect to personal features) and still break these principles in a critical situation. A transparent decision making process, however, requires and enforces congruence of principles and actions. The need for an explicit algorithmic implementation of ethical principles may thus point out a number of inconvenient questions.

---

[2]A more in-depth look into options for the regulation and certification of ethical decision making systems is given in appendix 6.8.

Furthermore, after any incident, the decision making system must allow to objectively trace and explicate why a specific behavior was triggered. The demand for transparency, therefore, not only concerns the public communication of the used approach, but requires the approach itself to be principally comprehensible by laypersons.

6.3.3 Public Consensus and Personalized Ethics Settings

Getting the ethical programming of an AV to reflect a public consensus may be difficult to achieve, as many people's intuitive values are at odds with existing legislation, or established legal precedence. A potential victim's age, for example, plays a major role to many people across all cultural backgrounds [5, 11, 39, 109, 112], but a differentiation by age or other individual factors is prohibited in most jurisdictions. For example, article 3 section 3 of the German constitution states "No person shall be favoured or disfavoured because of sex, parentage, race, language, homeland and origin, faith or religious or political opinions. No person shall be dis-favoured because of disability" [45]. Consequently, any attempt to differentiate or discriminate is strongly discouraged as well by the German ethics commission [88] (§3). Obviously, the programming of AVs must adhere to the constitutional rights of the respective jurisdiction.

Further complicating the goal of reflecting a public consensus in the AVs' programming is a significant variability in moral values even between individuals of the same culture [5, 112] (§4). One way to address this issue could be to use a parametric approach, aiming to express a person's moral values numerically. Such numerical representations could be assessed in forced choice decision tasks [5, 109], and the population's central tendency could be used in AVs to represent the most common ethical preferences of a society.

Alternatively, personalized ethics settings could allow the AV's owners to set up their cars to match their own ethical preferences, but this option is not without controversy [13, 49, 86, 12]. While much of this debate is focused on an optional over-protection of the AV's occupants, the option of allowing other ethical settings to be personalized has not gained much attention so far. A potential argument in favor of personalized ethics settings comes from research on forecasting algorithms: "Algorithm aversion" describes the tendency of people to choose human forecasters over algorithmic ones in all kinds of domains, even if they saw the algorithm perform significantly better than the human forecaster [32]. Dietvorst et al. [33] showed that algorithm aversion can be mitigated or avoided when people are given a chance to modify the algorithm's prediction, even if the changes they can make are very limited. Whether or not such findings translate to the field of automated ethics is an open, yet very interesting question. Allowing for limited personalization of certain ethics settings could potentially increase people's trust in, and satisfaction with algorithmic moral decisions, and might improve the adoption rates of AVs. While we are agnostic to the debate of personalized ethics settings vs. mandatory ethics settings at this point, it should be noted that the option of personalizing ethical preferences could be viewed as an advantage or even requirement for ethical decision making algorithms in the future.

6.4 Behavioral Principles in Dilemma Situations

The German ethics commission's report states important ethical rules for the behavior of automated vehicles in emergency situations, such as the protection of humans before all other considerations of utility, legal interests, and safety of animals and property [88]. It further states that "[a] set-off of victims is prohibited. A general programming to reduce the number of personal injuries can be justifiable. Those involved in the generation of mobility risks must not sacrifice uninvolved persons." The ethics commission thus appears to allow some form of utilitarian decision making, while insisting on a number of strict prohibitions stemming from a deontologically shaped understanding of morality. Unfortunately, most of these demands don't allow for a straightforward derivation of behavioral principles and are therefore unsuitable for a framework of automated ethical decision making. In the following, we will establish a set of important principles of ethical decision making in AVs, outline based on these where the aforementioned demands by the ethics commission are inadequate, and discuss alternatives.

6.4.1 Probabilistic Data and Robustness

The state of the vehicle and its immediate environment, as well as predicted future events will always be probabilistic in nature. The current state is reconstructed from sensor data, which always introduce some inaccuracies. The exact position and speed, for example of other cars or pedestrians, can only be estimated and will always carry some uncertainty. Even more so, future events can never be foreseen with absolute precision, and critical situations can be particularly difficult to predict. The behavior of other actors is often volatile, and the magnitude of a collision in terms of injuries can often only be coarsely approximated.

Thus, ethical decision making systems must be able to operate on probabilistic data (§5).

As a corollary of the probabilistic nature of the available data for decision making, the robustness of the ethical evaluations must be ensured. By robustness we mean that small changes in the situation should not lead to fundamental changes in the ethical evaluation. This is important for two reasons: First, ethical evaluations have to be repeated in short intervals. In an emergency situation, it is important that the ethical assessment doesn't jump back and forth between fundamentally different states, as this could cause volatile and unpredictable behavior by the vehicle. Second, if the ethical evaluations change fundamentally based on minimal differences between near identical situations, we lose transparency in the decision making process, violating the previously set demand. Thus, similar situations should lead to similar ethical evaluations (§6.1).

6.4.2 Categorical Distinctions, Thresholding, and Reasonable Decisions

The probabilistic nature of predictions implies that we often cannot know with certainty whether a person would die in a collision or only be injured. As a consequence, there is no straightforward solution to a categorical distinction between injuries and fatalities in the decision making logic. In the same way, we often cannot know whether or not a person would be injured in an accident, calling a categorical treatment of this variable into question as well. This becomes problematic when regulations or moral rules presume precisely these categorical distinctions, as is the case in the prohibition of all sacrifices of persons not involved for the greater good. Leaving aside for now the problematic conception of who is involved and who isn't, we have two choices: Either we refrain from all actions that pose the slightest risk of death to those not involved, or we define a threshold for the amount of risk above which the decision would be defined as a sacrifice – both of which are problematic.

Refraining from even the slightest risk is often incompatible with a notion of reasonableness. Let's say we need to swerve onto the sidewalk in order to save a pedestrian's life, and there is a 99% probability that no-one will be harmed at all, but a 1% chance that a second pedestrian on the sidewalk will be hit and killed in the process. We believe that it would be unreasonable not to take this small chance if it means saving another person with certainty. Yet if we deem even miniscule risks to bystanders unacceptable, this option is precluded from the start. The exact point at which it becomes unreasonable not to save the person at risk would need to be debated, but most will agree that at *some* point the absolute protection of bystanders from any risk will become unreasonable. We believe that a notion of reasonableness should be upheld whenever possible (§6.2).

Introducing thresholds that define at which level of risk a decision is considered a sacrifice could recover some notion of reasonableness, but only at the cost of robustness and transparency of the decisions. When the probability of a critical aspect is perceived near a defined threshold, smallest changes in the perception can fundamentally change the system's ethical evaluation of the situation, leading to the previously discussed lack of robustness. It would also be difficult to justify any precise threshold value. Most ethically relevant aspects of emergency situations exist on a continuous spectrum, and an artificial dichotomization would often not do them justice. Any precise threshold defining who counts as involved in a critical situation, or what constitutes a sacrifice, would be entirely arbitrary.

A possible solution to unite robustness of the decision making logic and reasonableness of the resulting decision would be to conceptualize ethically relevant properties on a continuous scale, and treat moral rules as soft constraints to the car's behavior. This would allow for a compromise between deontological and utilitarian considerations. The system would principally base its decisions on a comparison of the stakes involved for different parties in a situation, but could additionally disincentivise against the violation of important moral rules, as well as traffic violations. We, therefore, believe that an insistence on categorical distinctions and absolute prohibitions is not justified at the cost of reasonableness, robustness, and transparency in the decisions made. We do concede, however, that this is a possible point of contention and thus encourage the debate of this issue. Ultimately, a fundamental decision has to be made between insisting on strict categorizations and absolute prohibitions on the one hand, and making reasonable and robust decisions on the other (§7).

6.4.3 A Single Scale of Ethical Cost

The notion of reasonableness and the issues relating to categorical distinctions highlight another aspect that needs to be considered: Given that small or miniscule risks of human injury are often present in emergency situations and collision avoidance, is seems unreasonable to treat these as categorically superseding any risk to animals or material damages. For example, if a pedestrian steps in front of the car unexpectedly, and the car estimates a 10% chance of having a light collision with that pedestrian (speed at impact not exceeding 5 kph), should it take all risk from the pedestrian by swerving and crashing into a lamp post instead? Such behavior may be perceived by many as disproportionate. To prevent it, the well-being of humans would have to be set off with material and other ethical cost, such as the well-being of animals, at least so long as the risk of human injury is low. To assert that human well-being is still prioritized over all else, its valuation in the decision making logic would have to clearly exceed that of all other ethically relevant factors. The value of animals and material damages could also be capped to ensure that any severe or fatal injuries to humans supersede any amount of entirely materialistic damage or risk to animals. So long as severe injuries are very unlikely or precluded altogether, careful balancing of the involved cost factors could provide a solution that avoids unreasonable or disproportionate decisions. We thus believe it is justified to consider and debate the option of offsetting human well-being with other considerations on a single scale of ethical cost (§8).

6.4.4 Involvement, Responsibility, and Protection of AV Occupants

To address the question who should be considered involved in an emergency situation, and who would be protected as an innocent bystander, Birnbacher and Birnbacher [12] contend that from a moral point of view it is not relevant whether or not those in danger are in the movement direction of the car, since, unless otherwise specified, everyone in danger is equally innocent. The authors further argue that even if we allow a distinction between active "harming" and passive "letting harm occur", this distinction does not apply to AVs, since any behavior by the car would be active behavior, as pre-defined by the programmer. We agree with the authors in that an AV has no logically defined mode of "not behaving" while in motion, and all of its behavior can be regarded active. Those potentially in danger, however, arguably differ with respect to the amount of risk to themselves that they consent to, by means of their location and behavior. A pedestrian jaywalking generally consents to a larger amount of risk than a pedestrian crossing the street in an orderly manner, who, in turn, still consents to a greater level of risk than a person standing on the sidewalk. We suggest that a distinction be made with respect to the location of those who are potentially at risk (§9), i.e., the drivable surface could be divided into zones of differential protection to those within them. For example, sidewalks could be declared protected zones, drastically disincentivizing AVs from swerving in the direction of pedestrians.

With respect to assessing the responsibility of any involved party for causing a dangerous situation, there is an approach that lends itself to this possibility. The Responsibility-Sensitive Safety model (RSS) [103] is based on common-sense notions of accountability in traffic, and essentially formalizes a blame metric. While it was conceived to guide an AV's behavior such that it is never to blame for an accident, its assignments of blame could in principal be used to bring a notion of fairness into an AV's ethical evaluation of critical situations. While the technical and legal feasibility of such an approach would have to be examined in detail, and potential ethical issues would still have to be discussed, considering the amount of responsibility for a dangerous situation in the decision making logic could help to increase the perceived fairness of ethical decision making systems in AVs.

Either way, the general idea of differentiating by location and accountability of the involved parties could offer an elegant solution to another fiercely debated question [49, 13] – whether or not an AV should be allowed to provide special protection to its occupants at the cost of other road users. When other road users are getting protective advantages based on their location and possibly their (lack of) liability, one can discuss an appropriate protective classification of AV passengers. These are passive and thus never directly to blame for causing a dangerous situation. Yet by using the AV, they consent to some amount of mobility risk linked to the high speeds of the vehicle in comparison to pedestrians. Their protective classification would, therefore, sensibly lie between that of those at fault for causing the dangerous situation and that of bystanders on the sidewalk.

6.4.5 Breaking the Rules

Many scenarios in which smart collision avoidance and crash behavior can significantly improve the outcome of critical situations would require an AV to violate traffic regulations. Jumping a red light may be necessary to avoid being rear-ended, and swerving off the road, onto a sidewalk or into the opposite lane could sometimes prevent a head-on collision. We believe that it is unreasonable and even morally reprehensible not to avoid a collision if we know the risk to third parties to be minimal or non-existent.

Allowing for traffic violations would probably already be in line with the ethics commission's demands, which state that the protection of humans has to take precedence over other legal interests [88]. Beyond this, it could be viewed as sensible to allow for traffic violations also to prevent material damages, if no considerable risk to others is created in the process. Similar to moral rules, traffic regulations could be implemented as soft constraints, assigning a cost to certain violations that can be outweighed by defined amounts of risk-minimization.

Ultimately, traffic regulations serve the purpose of improving safety, and not standing in its way. We thus argue that responsibly breaking traffic regulations to benefit overall safety is not only acceptable, but a moral imperative when the circumstances demand it. AVs should thus be allowed or even encouraged by regulation to break certain rules if the car is equipped to make the respective decisions *sensibly* (§10).

6.4.6 Synopsis

In summary, the above discussion shows a number of severe limitations of the guidelines by the German ethics commission [88], and gives rise to the following list of demands and open questions that need to be addressed in the pursuit of a solid framework for algorithmic ethical decision making in AVs:

1. An AVs ethical behavior needs to be regulated. A precise definition of acceptable behavioral principles for AVs in critical situations is necessary, and a verification that these principles are upheld by the software needs to be part of the certification procedure for AVs.

2. The behavioral principles implemented in decision making systems of AVs must be transparent, traceable, publicly communicated, and be principally comprehensible by laypersons, precluding black-box algorithms.

3. Personal characteristics of potentially involved persons, such as their age or gender, must not be considered in the decision making procedure.

4. It needs to be determined, to which degree behavioral principles in dilemma situations can and should reflect a public consensus, and whether or not personalized ethics settings, possibly restricted to a defined range, are an option.

5. All behavioral rules and systems that implement these must do justice to uncertainties in perception and the probabilistic nature of predictions.

6. Ethical evaluations should be robust, i.e., similar situations should result in similar ethical evaluations, and a notion of reasonableness in the decisions should be upheld whenever possible.

7. It needs to be debated, to what extent an insistence on categorical distinctions with absolute prohibitions can be reconciled with

robust, transparent, and reasonable decisions in a probabilistic framework.

8. The protection of human life needs to take priority over that of animals or property. It should be discussed if and to what extent a set-off of human well-being with other ethically relevant factors is acceptable when it serves to avoid disproportionate decisions.

9. In the decision making logic, a distinction needs to be made with respect to the location of a person at risk. It should also be debated, whether the amount of responsibility for the dangerous situation can or should be factored in as well.

10. AVs must be allowed to violate certain traffic regulations if the car is equipped to make the respective decisions sensibly.

6.5 Approaches Towards a Solution

With respect to concrete solutions, some existing approaches generally comply with the stated demands. van Otterlo [122] proposes a utility function accumulating all costs associated with each trajectory option, and selecting the one with the lowest cost. The costs per trajectory option are a sum of the "ethical costs" per killed obstacle and other consequences of choosing a given trajectory. However, this approach does not explicitly consider the probabilistic basis for the decisions. Addressing this aspect, Goodall [52] suggests that AVs could perform risk analysis by calculating expectation values, i.e., the product of the probability of an unwanted event, and its magnitude. An event with a 10% chance of occurring, which would cost two lives would thus have an expectation value of 0.2. The process of balancing multiple sources of risk, or multiple expectation values, called risk management, would then be the basis for the decision making process. We believe that a combination of utility functions and risk management could be conceived that fulfills all of the previously established framework conditions. However, any concrete approach would have to pick a side on the open questions outlined above.

6.5.1 Examples of Dilemma Situations

To provide a more illustrative account of some of the issues outlined above, we will exemplify them with two hypothetical scenarios.

Scenario 1

An AV is driving in city traffic with two passengers on board, doing the allowed 50 kph. The road has a single lane in each direction and sidewalk to the right of the car. Without checking for traffic, a pedestrian (pedestrian 1) starts crossing the road a few meters in front of the car, and the AV is left with three options:

1. Initiate an emergency breaking maneuver without swerving. The car's perception module estimates an 80% chance of a collision, and the collision would likely kill the pedestrian. The AV's passenger would not be harmed.

2. Swerve on the sidewalk to the right and initiate emergency breaking. A second pedestrian (pedestrian 2) is on the sidewalk a little further away. The perception module estimates a 70% chance of a collision with the second pedestrian. Due to the increased distance, and thus reduced speed at impact, the collision is estimated to cause severe injuries, but unlikely to kill them. The AV's passenger would not be harmed.

3. Swerve to the left into oncoming traffic and initiate emergency breaking. A car with a single driver and no further passengers would be hit with near certainty (95%), likely resulting in minor injuries for all passengers in the two cars.

Scenario 2

An AV with one passenger on board is stopped at a red light on a suburban intersection. The car behind the AV with one person on board initiates a late emergency braking maneuver, and a rear-end collision is probable. The intersection is empty with one car approaching from the left (two occupants). This car has a green light, but would probably be able to swerve or come to a stop in time if necessary. The AV has the following options:

1. Stay put with a 70% chance of being rear-ended. The AV predicts minor injuries to passengers of both vehicles if the collision happens, while the car approaching from the left would be unaffected.

2. Jump the red light to give the car behind enough space to come to a stop. The AV calculates a 10% chance of colliding with the car approaching from the left, which would likely result in minor injuries for the other car's passengers, and moderate injuries to the AV's passenger.

These examples illustrate the probabilistic nature of the situations
in which the AV needs to make a decision, and show that a large number of different factors need to be evaluated at the same time. These
include the well-being of multiple parties, material damages, traffic regulations, and ethical considerations. It is easy to further elaborate these
scenarios, e.g., involving animals, persons of different age, or unknown
numbers of passengers in a car or bus. This shows that the complexity
of real world decisions more often than not touches upon a majority of
the above ten requirements and open questions.

6.6 Trajectory Choices

Finding appropriate solutions to these issues would go a long way in
creating a framework for the regulation and implementation of ethical
decision making systems. However, two important practical considerations have not yet been addressed: (1) Trajectory-selection in continuous space instead of a low number of distinct options, and (2) the option
of optimizing trajectories in regular driving.

6.6.1 Evaluations of Continuous Space

In most examples of traffic dilemmas, the scenarios are presented in a
trolley-like fashion with typically two or three very distinct behavioral
options for the driver or vehicle. While this makes it easier for humans
to get a grasp of the situation and discuss it, this view may be rather
detached from how path planning and decision making works in reality.
Between every two distinct behavioral options, there is technically an
infinite number of trajectories the car could take, and it is unclear how
it should decide, which of these to evaluate in detail. Thus, the problem
is not only how to decide, but already which options to investigate for
a later decision.

Instead of evaluating a low number of trajectories representing fundamentally different choices, one could integrate the ethical evaluation
with the car's internal map of its immediate surrounding to achieve an
ethical evaluation of continuous space. This map, in Gruyer et al. [60]
referred to as the local dynamic perception map (LDPM), represents the
vehicle's position and dynamic state, lanes and surfaces, as well as possible obstacles and their predicted trajectories[3]. Based on the LDPM, a
map of the car's environment could be created, in which each point in

---

[3]A more detailed account of the perception and control architecture in AVs is given
in appendix 6.8.

space (or pixel) is assigned a value that answers the question "how good or bad would it be to choose a trajectory that leads the car over this point in space?". Figure 6.1 gives an intuition of how such an approach could work: (A) is a depiction of the scene in the real world, with the blue car being the AV. In its perception module it detects a number of features of its surrounding scene, such as relevant zones (lanes, walkways, etc.), depicted in (B), and relevant objects (pedestrians, other cars, etc.), depicted in (C). Since many of the relevant objects in (C) are moving, a prediction of their future locations is computed and indicated as probability densities. In the ethical evaluation step, the detected zones and locations of relevant objects would be transformed into disincentives to move into a given space. (D) shows a small disincentive to move into the opposite lane, and moderate disincentives to move onto the sidewalk. (E) shows strong disincentives to move into spaces predicted to be occupied by other objects, in particular by pedestrians on the sidewalk. Finally, (F) is a superimposition of D, and E, and could be used to precisely plan the vehicle's trajectory. Note that the illustration doesn't cover all contents of an LDPM, nor all aspects relevant to the ethical evaluation of the scene, and disregards the passing of time for simplicity.

6.6.2 Trajectory Optimization in Regular Driving

An ethical evaluation of continuous spaces would pave the way for yet another option: Optimizing the vehicle's trajectory not just in critical situations, but permanently. Since there are no precise boundaries between careful maneuvering, collision avoidance, and crash behavior, defining a precise switch point at which the ethical decision making system takes over would appear arbitrary and might leave safety potential unused. Instead, a constant ethical assessment for the space around them would allow for a sensible adaptation of their driving speed, let them assess when it is safe enough to overtake a slower vehicle or a cyclist, and position themselves in their lane in a way that always minimizes the risk of a collision.

Even in the absence of any immediate and recognized hazards, occluded areas, such as the space behind parked vehicles on the side of the road, could be considered individual sources of risk, creating small disincentives against driving past them at a very close distance. Under certain circumstances, their risk cost could be treated as elevated, for example when other parts of the sidewalk are visibly crowded with pedestrians. In a narrow street with parked cars on the side but no other traffic, the car would then move closer to the center in order to leave

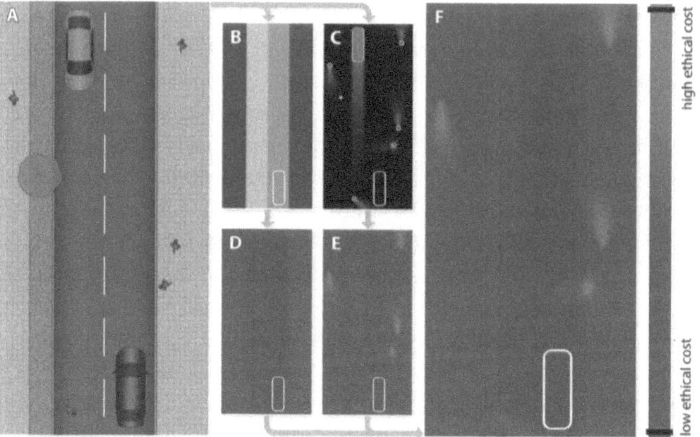

Figure 6.1: Sketch: Ethical evaluations in continuous space. A: Depiction of a scene in the real world. B: Detected surfaces (road, lanes, sidewalks, etc.) as part of the LDPM. C: Detected objects, color-coded by type, and corresponding movement prediction (probability densities) as part of the LDPM. D: Ethical evaluation of drivable surfaces: Deeper reds indicate higher ethical cost (stronger disincentives) to drive in that region. E: Ethical evaluation of objects based on their predicted locations, i.e., probability densities weighted by vulnerability and location. F: Full ethical evaluation of the scene (superimposition of D and E).

more room for possibly occluded pedestrians, and more time for itself to avoid a collision.

However, since very small amounts of risk are present in almost any situation, the decision making system would have to go beyond a mere minimization of risk to a process of counterbalancing risk and utility. Without accounting for utility, the system would often be likely to stop the car immediately, since not moving at all is the safest option in most cases. The aspects of utility that would be considered could include economy of time – creating an incentive to keep the car moving –, passenger comfort, energy consumption and conformity with traffic culture. The full integration of these aspects of utility could lead to ethical decision making systems covering every aspect of driving.

6.7 Conclusion

In this work, we analyzed when and why ethical decision making systems become necessary in AVs, we reviewed existing guidelines for the behavior of AVs in dilemma situations, and we compiled a set of 10 de-

mands and open questions that need to be addressed in the pursuit of
a framework for ethical decision making in AVs. The principal issues
outlined in this work show that neither the regulations for algorithmic
ethical decisions in AVs, nor a practical implementation of such a system
can be conceived independently of one another. Any tangible solution
for automated ethics needs to reconcile fundamental law, technical feasibility, and the moral values of the society. The goal must be to not
only allow the cars to be as fair and safe as possible, and guard the
manufacturers from legal liability with respect to the ethical programming of their cars, but also to foster public trust in automated driving
technology, and facilitate its adoption. It is, therefore, important that
regulatory bodies and the industry engage in a dialogue to find an appropriate solution. At the same time, the central ethical issues need to
be discussed publicly, and we emphatically encourage this debate.

Chapter 7

Adaptive Blending Units: Trainable Activation Functions for Deep Neural Networks

Contributions

ArXiv.org: Leon René Sütfeld, Flemming Brieger, Holger Finger, Sonja Füllhase, and Gordon Pipa 2018

LRS: Conceptualization, Programming, Methodology, Project administration, Data curation, Formal analysis, Visualization, Writing – original draft; FB: Programming, Methodology, Investigation, Methodology, Writing – review & editing; HF: Conceptualization, Investigation, Methodology, Writing – review & editing; SF: Conducting Pilot Study, Writing – review & editing;; PK: Supervision, Writing – review & editing; GP: Supervision, Writing – review & editing

Layman's Summary

In practice, good ethical decisions require an accurate comprehension of the situation around the vehicle. The predominant technology used to process the data from cameras and other sensors for the purpose of scene and entity recognition is machine learning, and in particular deep neural networks. The final publication of this theses, therefore, looks into advances in deep learning.

Giving deep neural networks their name is the computational principle involved: Computational nodes ("neurons") are organized in layers, which are interconnected and stacked on top of each other. The use of a large number of layers (thus, "deep" networks) allows these networks to perform complex computations on the input. Each layer consists of a linear transformation of the previous layer's output, defined in a weight matrix, and a non-linear transformation, defined in the networks so-called activation function. The repeated use of non-linear transformations gives the networks their computational power, and thus activation functions play a central roll in their success. But while the weight matrices are learnt during training of the network, the activation function is typically chosen beforehand and kept fixed for both training and inference.

To date it is still unclear what properties of an activation function are most desirable, given a certain task and network specification. We approach this issue by introducing the Adaptive Blending Unit (ABU), a trainable activation function that is defined as a weighted sum of a number of conventional activation functions. The weights, and thus the shape and scaling of the ABU are learnt during training alongside the network's weight matrices, and fixed during inference. We experimentally compare the performance with the highest performing conventional activation functions over a variety of network architectures, settings, and tasks. Our results show ABUs to be the highest performing activation function averaged across the chosen tasks and network settings.

In our analyses, we further separate the effects of shape and scaling of the activation function, providing valuable insight into their impact over the course of training. As these findings apply universally to various types of deep neural networks, we chose not to discuss specific areas of application, such as the use in automated vehicles, in the paper.

7.1 Introduction

Deep neural networks are structured around layers, each of which performs a linear transformation of its input before feeding the signal through a scalar non-linear activation function. Chaining larger numbers of non-linear functions then allows the networks to find and extract complex features in the input. Activation functions thus have a key function in deep neural networks: Without intermittent non-linearities, these networks could only perform linear operations on the input. But despite a large number of activation functions proven successful in the literature, it remains unclear, what properties of an activation function are most desirable, given a particular task and network configuration. Ideally, the network would sort this issue out by itself, but most common activation functions are fixed during training, i.e., their shape and scaling are treated as hyperparameters. We suggest changing this practice by making an activation function's shape and scaling a trainable parameter of the network. Our main contribution in this work is the Adaptive Blending Unit (ABU), a linear combination of a set of basic activation functions that allows the shape and scaling of the resulting activation function to be learned during training. In an effort to separate and understand the effects of the activation function's shape and its scaling, we also examine the effect of adaptive scaling on common activation functions without adaptation of the shape, as well as normalizing the blending weights in ABUs, thus learning its shape without learning any scaling. Throughout this work, we apply one scaling weight, or one set of blending weights (i.e., one ABU) per layer of the network. This way, the network is free to learn the activation function and / or scaling that best suits the computations performed in any given layer, while the number of parameters in the network is kept low enough, as not to require regularization. The remainder of this work is structured as follows. In section 2, we will review related work, before comparing and analyzing common activation functions, their adaptively scaled counterparts and ABUs on CIFAR image classification tasks in section 3. In section 4, we examine multiple ways of normalizing ABUs, to provide an account of adaptive shape without adaptive scaling. Finally, in section 5, we examine pre-training of the scaling and blending weights, to examine the role of adaptiveness over the course of training. We conclude the paper by discussing limitations of the chosen approach, and providing an outlook on future work on this topic.

7.2   Related Work

The most prevalent activation function in modern neural networks is the *Rectified Linear Unit (ReLU)* [61, 94], a piecewise-linear function returning the identity for all positive inputs and zero otherwise. Its constant derivative of 1 on the positive part helps alleviating the vanishing gradient problem [46], making it the first activation function allowing for a large number of stacked layers to be trained efficiently. With this, ReLU was partly responsible for the breakthrough of deep neural networks around 2012, marked by AlexNet's victory in the annual ILSVRC challenge [81]. *Leaky ReLU (LReLU)* [89], *Parametric ReLU (PReLU)* [63], and *Randomized Leaky ReLU (RReLU)* [127] are all based on ReLU, but replace the zero-output for negative values by a linear function. In PReLU, the slope of the negative part of the function is controlled by a trainable parameter. Exponential Linear Units (ELU) [22] like ReLU, return the identity for positive values, but $\alpha(\exp(x)-1)$ for negative values, with α typically set to 1. Scaled Exponential Linear Units (SELU) [77] are identical to ELU, except for an additional scaling parameter λ acting upon the function as a whole. The values for α and λ in SELUs are analytically derived to ensure convergence of activations towards unit mean and variance across layers. In a more empirical approach, [99] performed a large reinforcement learning-based search for successful activation functions, and found multiple novel and well-performing functions. The most successful, given by $f(x) = x \cdot \text{sigmoid}(\beta x)$ and named *Swish*, uses the trainable parameter β to control the overall shape of the function. E-Swish [2] ditches this parameter (setting it to 1), and instead scales the whole function by a manually determined parameter between 1 and 2. In addition to these, there are numerous approaches in which the activation function's overall shape is learned, often using multiple parameters: *Adaptive Piecewise Linear Units (APL)* learn the slope of all piecewise linear elements and the position of the hinges independently for each neuron via backpropagation, while the number of linear pieces is a hyperparameter that is set manually [1]. Similarly, Maxout activations [53] learn a convex piecewise-linear function by returning the maximum of a fixed set of neurons, while the regular network weights determine the shape of the resulting function. [36] use Fourier series basis expansion to approximate non-linear parameterized basis functions, and train one activation function per feature map in a convolutional network. An approach suggested by [47], called the *soft exponential function*, can switch between a large number of different mathematical operations, such as addition, multiplication and exponentiation, by adjusting a trainable parameter. However, to our knowledge, no empirical validation of the approach was offered so far. In an approach similar to ours, [35] suggested blending a set of activation functions on a per-neuron basis, and constraining the blending weights to values between 0 and 1, by gating them with exponential sigmoid functions. Blending activation functions on a per-neuron basis, however, required downscaling of gradient updates as a form of regularization. Most similar to our approach, [90] suggested a learned blending of multiple common activation functions per layer, where the blending weights are constrained to sum up to 1, and showed this approach to be successful over a range of tasks and network configurations. We will provide further details with respect to the similarities between their approach and ours in the appropriate sections.

7.3 Adaptive scaling and Adaptive Blending Units

In this section, we will introduce ABUs as an extension to the idea of an adaptive scaling of common activation functions, and analyze both ABUs and adaptive scaling with respect to task performance and the mechanics they introduce to the network. The activation functions we used as a baseline throughout this work are the *hyperbolic tangent (tanh)*, *ReLU, ELU, SELU*, the *identity* and *Swish*. We will reference the adaptively scaled versions of these by adding "α" to the function's name, e.g., "αReLU".

7.3.1 Methods

Given a deep neural network of n layers, and an activation function $f(x)$, the adaptively scaled version of the activation function is given by $\alpha_i \cdot f(x)$, with $i = 1, \ldots, n$. The *scaling weights* α_i are initialized at 1 by default, and trained via backpropagation alongside all other network parameters. Swish's β is initialized as a trainable parameter per layer (i.e., β_i) and likewise trained via backpropagation in all cases. ABUs can be viewed as an extension to this approach, in which the shape of the activation function is determined by a blending of multiple common activation functions within the unit. Given a deep neural network of n layers, and a set of m activation functions per layer, the ABU for the ith layer is defined as $g_i(x) = \sum_j \alpha_{ij} \cdot f_j(x)$ with $i = 1, \ldots, n$ and $j = 1, \ldots, m$. The *blending weights* α_{ij} are initialized at $\frac{1}{m}$ by default, and also trained via backpropagation alongside all other network parameters. With respect to the set of activation functions used in ABUs,

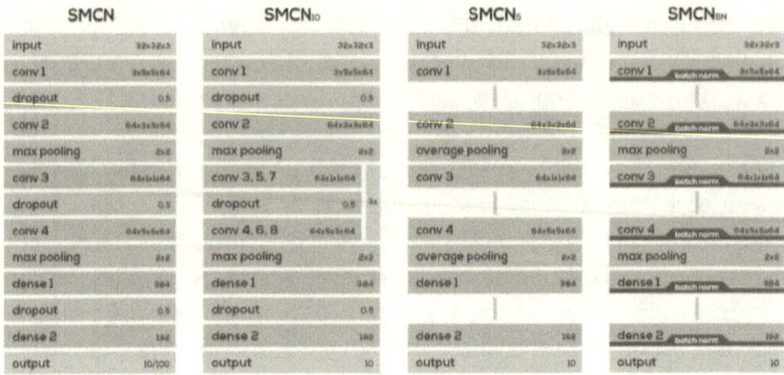

Figure 7.1: SMCN network architectures for CIFAR10 / CIFAR100. **(Vanilla) SMCN**: Default architecture, 6 hidden layers. **SMCN$_{10}$**: Medium-sized network, 10 hidden layers. **SMCN$_S$**: Additional architecture for robustness tests; 6 hidden layers, omitting dropout layers and replacing max pooling with average pooling. **SMCN$_{BN}$**: Additional architecture for robustness tests; 6 hidden layers, batch normalization after each activation, and omitting dropout layers.

we chose tanh, ELU, ReLU, the identity, and Swish in order to allow for high flexibility of the resulting function. However, we did not conduct an exhaustive search over possible sets of activation functions, so other sets may outperform the chosen configuration.

The *CIFAR 10* and *CIFAR 100* datasets [80] served as benchmarks to assess the performance of our approaches. Per-image z-transformation was applied as pre-processing to all images, and 5% of the training set was used as a validation set during training. To evaluate model performance, we applied post-hoc early stopping: The model was saved once every 8 epochs and the validation accuracy was estimated frequently over the course of training. All networks were trained for 60 000 steps, after which we smoothed the validation accuracy curve and selected the model save point for which said curve indicated the highest performance. For each network and task specification, we report the mean of 30 runs, as well as the standard error. Training, validation and testing were all performed using mini-batches.

For the networks used in our tests, we created a set of small to mid-sized convolutional networks, called Simple Modular Convolutional Networks (SMCN). In different variations of these, features were added or subtracted to test the robustness of our approaches across different network design choices. The vanilla SMCN consists of four convolutional layers, followed by two dense layers. Max pooling (3×3, stride 2) is performed after the second and fourth convolutional layer, and dropout [106] with a rate of 0.5 is applied after the first and third convolutional layer, and after the first dense layer. The convolutional layers use zero-padding and stride 1. They feature filters of size $[5 \times 5 \times 64]$, $[3 \times 3 \times 64]$, $[1 \times 1 \times 64]$, and $[5 \times 5 \times 64]$, and the dense layers consist of 384 and 192 neurons, respectively (see Figure 7.1). The network contains no residual connections, and no batch normalization [69] is performed by default. Initial weights are randomly sampled using He initialization [63]. Bias units were initialized at 0.1, except for the first convolutional layer (0.0). The network is trained for 60 000 steps on mini-batches of size 256, using the Adam optimizer [76] with a learning rate of $\eta = 0.001$. The total number of trainable parameters in the vanilla SMCN is roughly 1.8M. In addition to this, we used the following variations in our tests: $SMCN_{10}$, a mid-sized network (10 layers, roughly 2.0M parameters) identical to SMCN, with the exception that all layers between the two max pooling operations are repeated three times. $SMCN_S$, a simplified architecture where max pooling was replaced with average pooling, and all dropout layers were removed from the network (thus, the activation functions constitute the only non-linearities in this network). $SMCN_{BN}$, in which batch normalization is performed before applying the activation functions. We decided not to use dropout in this architecture, as batch normalization in conjunction with dropout can be problematic [84]. Note that since batch normalization negates the effect of any preceding scaling, adaptive scaling should not make a difference here. Finally, we also tested the vanilla SMCN with a Stochastic Gradient Descent optimizer with Momentum, instead of the Adam optimizer. Here, the networks were again trained for 60 000 steps, with the momentum parameter set to 0.9, an initial learning rate of $\eta = 0.01$, and a learning schedule linearly decreasing the learning rate per weight update, reaching 0.0004 at the end of training.

7.3.2 Performance

For our performance tests, we chose a vanilla SMCN with Adam optimizer and CIFAR10 as the default setup to compare the various activation functions. All other tested setups are systematically varied versions of this, and differ in only one aspect each, i.e., network architecture, optimizer, or task. On average, adding adaptive scaling yielded improved performance for all activation functions, as evidenced by higher mean ranks of all adaptively scaled activation functions, compared to their fixed counterparts (see Table 7.1). However, as expected beforehand, batch-normalized networks ($SMCN_{BN}$) were found to be indifferent to adaptive scaling. Interestingly, also in networks trained with the Mo-

Table 7.1: Performance comparison (percentage correct): Common activation functions, adaptive scaling, and ABUs by task, optimizer and network architecture. Table shows mean values of 30 runs plus standard errors per configuration, as well as mean rank across all six configurations. Highest performing activation function per column in bold.

| Network | SMCN | SMCN_{10} | SMCN_s | SMCN_BN | SMCN | SMCN | |
| Optimizer | Adam | Adam | Adam | Adam | Momentum | Adam | Mean |
Task	CIFAR10	CIFAR10	CIFAR10	CIFAR10	CIFAR10	CIFAR100	Rank
I	75.51 ± 0.11	73.19 ± 0.37	38.87 ± 0.08	71.72 ± 0.09	77.34 ± 0.06	44.11 ± 0.09	12.00
αI	76.52 ± 0.08	77.34 ± 0.18	39.48 ± 0.06	71.34 ± 0.10	76.32 ± 0.06	45.58 ± 0.09	11.50
tanh	75.44 ± 0.06	58.55 ± 4.47	67.19 ± 0.11	75.10 ± 0.07	78.76 ± 0.05	41.02 ± 0.13	12.00
αtanh	79.07 ± 0.07	73.40 ± 3.87	68.82 ± 0.10	75.32 ± 0.10	79.14 ± 0.05	46.85 ± 0.08	9.83
ReLU	79.42 ± 0.17	81.07 ± 0.15	72.79 ± 0.16	81.17 ± 0.06	81.63 ± 0.07	43.66 ± 0.10	8.17
αReLU	79.23 ± 0.15	82.97 ± 0.12	73.89 ± 0.10	81.12 ± 0.11	81.85 ± 0.07	46.22 ± 0.11	7.17
ELU	81.78 ± 0.06	83.41 ± 0.08	73.33 ± 0.13	80.87 ± 0.06	82.16 ± 0.06	48.59 ± 0.11	5.83
αELU	82.60 ± 0.06	84.94 ± 0.06	75.03 ± 0.13	80.89 ± 0.06	82.06 ± 0.04	51.03 ± 0.10	3.50
SELU	81.75 ± 0.07	83.29 ± 0.07	71.72 ± 0.14	79.36 ± 0.05	82.48 ± 0.05	48.25 ± 0.08	6.83
αSELU	82.81 ± 0.06	$\mathbf{85.04 \pm 0.04}$	73.79 ± 0.15	79.57 ± 0.07	81.99 ± 0.05	51.08 ± 0.08	4.33
Swish	82.07 ± 0.08	83.73 ± 0.07	74.33 ± 0.16	$\mathbf{81.77 \pm 0.04}$	82.02 ± 0.06	49.14 ± 0.12	4.33
αSwish	82.27 ± 0.06	84.56 ± 0.05	75.67 ± 0.09	81.61 ± 0.05	82.35 ± 0.05	50.19 ± 0.08	3.17
ABU (ours)	$\mathbf{83.12 \pm 0.06}$	84.70 ± 0.06	76.19 ± 0.11	80.63 ± 0.09	$\mathbf{83.12 \pm 0.06}$	$\mathbf{52.13 \pm 0.08}$	2.33

mentum optimizer, adaptive scaling yielded little to no improvement over the fixed activation functions. Adaptive Blending Units, on the other hand, outperformed all other activation functions in four out of six setups (including the Momentum setup), showing remarkable robustness across architectural choices, and consequently scoring the highest mean rank of all tested activation functions. Since the ability to perform adaptive scaling is an integral part of Adaptive Blending Units, any improvements over adaptively scaled activation functions can likely be attributed to their adaptive shape.

7.3.3 Analysis

But what exactly changes in the networks, when we introduce adaptive scaling or ABUs? In order to provide some insight into the mechanisms introduced by the two approaches, we carried out further analyses based on the default setup, i.e., a vanilla SMCN with Adam optimizer trained on CIFAR10.

Let us first examine how scaling weights behave during training. In our tests, the scaling weights α_i almost unanimously decreased to values far below 1 (see Figure 7.2A). This behavior was highly consistent over repeated runs with random initializations and mini-batch sampling: The mean standard deviation (over repeated runs) of the final scaling weights reached at the end of training is $\text{mean}(\sigma_{\alpha_i}) = 0.023$. With respect to how this influences the activations in the network, it is sensible to consider the succeeding layer's pre-activation statistics, i.e., the distribution of values going into its activation function: The distri-

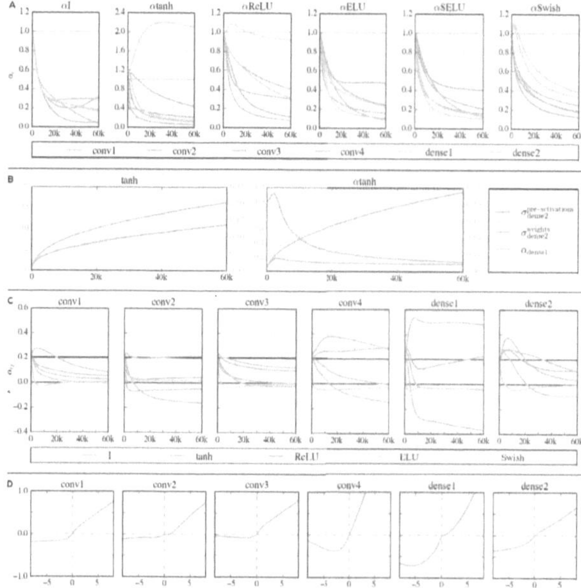

Figure 7.2: **A**: α_i of adaptively scaled activation functions over the course of training in a vanilla SMCN (mean of 30 runs, 60 000 steps). **B**: Effect of adaptive scaling on pre-activation distributions, exemplified by tanh & αtanh. Scaling weights α_i (magenta) enforce stable variance of next layer's pre-activations (orange), compensating for the increased variance of the regular weight matrix W_{dense2} (blue). Plots show mean of five runs. **C**: ABU blending weights α_{ij} over the course of training. **D**: Average activation functions (ABU) by layer at the end of training (axes scaled to improve readability).

bution of pre-activation is approximately Gaussian for large layers due to the Central Limit Theorem, and is thus easier to compare between networks with different activation functions. For many activation functions, the pre-activations are also crucial with respect to the magnitude of the gradients, in that they determine the fraction of inputs reaching saturated regions of the activation function. Our analyses show that the decreasing scaling weights rather precisely counteract an increase in the variance of the following weight matrix over the course of training. This stabilizes the distributions of pre-activation states in the following layers in both mean and variance, thus drastically reducing any covariate shift. We illustrate this by comparing the pre-activation variance of the last layer in SMCN networks, using tanh and αtanh, in Figure 7.2B. Without adaptive scaling, the variance of pre-activations increased throughout training for all layers and all tested activation functions. With adaptive scaling, the standard deviations typically converged to a value between 0.5 and 5 early on in training, and remained stable at this value for the remainder of the training procedure. At the same time, pre-activation means were kept stable at less than a standard deviation from zero.

We take from this that adaptive scaling acts as a normalization technique, similar to batch normalization [69] or layer normalization [6]. In contrast to these, however, adaptive scaling doesn't require any explicit calculations of variance or other statistics, or keeping track of running averages in inference, and does not depend on batch or layer size. In principle, it also allows the network to optimize the statistics of the neurons' input distributions, instead of enforcing unit mean and variance across the layer or batch. Depending on the type of calculations that are predominantly performed by a given layer, it appears plausible that some layers would prefer stronger saturation, while others may benefit from less saturation, provided the activation functions feature increasingly saturating regions. That being said, our analysis does not allow us to infer whether or not the realized distributions actually constitute an optimum for the required computation in a given layer. If an activation function is a homogeneous function of degree 1 (scale-invariant; e.g., ReLU), the network performance would likely not be influenced by the variance of the pre-activation's distribution, but may still be affected by shifts of the mean, which are also mitigated by adaptive scaling. We consider an in-depth analysis of such self-organizing processes, as well as further exploration of this principle for deep networks highly desirable, but out of scope for this work.

Turning to ABUs, we observe the same normalizing effect on the pre-activation statistics of succeeding layers. As illustrated in Figure 7.2C, ABUs realize a layer's overall downscaling in multiple ways. In the first convolutional layer, for instance, the weights unanimously decrease and mostly converge towards values close to zero. At the end of training, the identity and ReLU have arrived at effectively zero, while the final activation function is mostly a mixture of ELU and tanh. By contrast, the first dense layer achieves the overall downscaling of positive inputs by subtracting ReLU from a mixture of ELU and Swish. In both cases, the resulting function is rather flat, pushing the activations (layer output) closer to zero. These different compositions of blending weights translate into substantially different shapes of the resulting ABU (see Figure 7.2D). But while the variation of the ABUs' shape across layers is substantial, their shape within each layer is remarkably consistent over repeated runs, as indicated by a mean standard deviation of

Table 7.2: Performance comparison (percentage correct): ABU and the normalized ABU_{abs}, ABU_{nrm} ABU_{pos}, and ABU_{soft} by task, optimizer and network architecture. Table shows mean values of 30 runs plus standard errors per configuration, as well as mean rank across all six configurations. Highest performing activation function per column in bold.

| Network | SMCN | $SMCN_{10}$ | $SMCN_S$ | $SMCN_{BN}$ | SMCN | SMCN | |
| Optimizer | Adam | Adam | Adam | Adam | Momentum | Adam | Mean |
Task	CIFAR10	CIFAR10	CIFAR10	CIFAR10	CIFAR10	CIFAR100	Rank
ABU	83.12 ± 0.06	84.70 ± 0.06	$\mathbf{76.19 \pm 0.11}$	80.63 ± 0.09	83.12 ± 0.06	$\mathbf{52.17 \pm 0.08}$	2.5
ABU_{nrm}	82.82 ± 0.07	84.18 ± 0.07	75.99 ± 0.10	81.39 ± 0.07	$\mathbf{83.29 \pm 0.06}$	51.88 ± 0.10	3.7
ABU_{abs}	$\mathbf{83.14 \pm 0.08}$	$\mathbf{84.95 \pm 0.06}$	76.07 ± 0.16	81.17 ± 0.08	82.32 ± 0.05	52.16 ± 0.07	2.7
ABU_{pos}	82.90 ± 0.05	84.05 ± 0.06	76.10 ± 0.11	81.44 ± 0.07	83.26 ± 0.06	51.90 ± 0.09	3.2
ABU_{soft}	82.54 ± 0.07	84.63 ± 0.05	76.18 ± 0.07	$\mathbf{81.51 \pm 0.06}$	82.09 ± 0.05	52.10 ± 0.08	3.0

$\sigma_{mean}^{\alpha_{ij}} = 0.010$ per layer and blending weight. This consistency, in conjunction with the good performance figures achieved by ABUs, lead to the conclusion that the learned shapes are meaningful with respect to the computations performed in the network.

In summary, while adaptive scaling stabilizes the pre-activation statistics of succeeding layers, the learned shapes of the resulting functions are meaningful, as well. Moreover, both adaptive scaling and an adaptive shape were found to yield improvements in performance for image classification tasks with convolutional networks.

7.4 Normalized Blending Weights

So far, we focused on adaptive scaling as an integral part of ABUs. In order to better understand the effects of shape in ABUs, we conducted an additional experiment, in which we normalized the blending weights of the ABUs in four different ways, taking away their ability to scale the layer output by overall increases or decreases of the blending weights.

7.4.1 Methods

The following four methods of normalization for ABUs were used: ABU_{nrm} denotes the case where a layer's blending weights are normalized to sum up to 1 ($\sum_j \alpha_{ij} = 1$). The normalization was implemented as part of the graph, dividing the blending weights by their sum, before applying them in the respective ABU. Similarly, in ABU_{abs}, we divided the raw blending weights by the sum of their absolute values, thus keeping the sum of the absolute values of the blending weights at 1 ($\sum_j |\alpha_{ij}| = 1$). Note that under this constraint, scaling is still possible, albeit not independent of the resulting shape: By having similar activation functions cancel each other out with blending weights on either side of zero, functions can be constructed that return only a fraction of the input, or even zero, for all positive values. We decided to include this form of normalization in the test to provide a more complete account of possible normalizations. In ABU_{pos}, any negative values are clipped before normalization, such that all blending weights are strictly positive ($\sum_j \alpha_{ij} = 1$; $\alpha_{ij} > 0$). Finally, in ABU_{soft}, we realized the same constraint (all-positive and normalized) by applying softmax normalization to the blending weights. With the exception of ABU_{abs}, none of the normalized versions of ABUs can realize an overall scaling of the resulting functions for positive input values. For the experiments, we used the same network and task configurations as in the previous section.

7.4.2   Performance & Analysis

The results of our performance tests are reported in Table 7.2. All five versions of ABUs showed remarkably similar performance throughout the tested task and network configurations - the average gap between the best and weakest performing ABU in a given setup is a mere 0.63%. In terms of mean rank, ABU and ABU_{abs} lead the field, and are thus the two most robust configurations. However, none of the other three versions fell far behind. We again used the default setup (vanilla SMCN, Adam, CIFAR10) for an analysis of the blending weights and their effects on the succeeding pre-activations. We found ABU_{abs}s to behave much like unconstrained ABUs, implementing adaptive scaling, keeping the layer statistics constant over the course of training, thus mitigating covariate shift. Despite the fact that the scaling imposes constraints on the shape of the resulting function (as outlined above), ABU_{abs} performed very similarly to unconstrained ABUs in most settings. By contrast, but very much expectedly, ABU_{nrm}, ABU_{pos}, and ABU_{soft}, were unable to keep the layer statistics at constant levels, and a considerable covariate shift akin to that in fixed activation functions was observed. Interestingly, this appears to have only a minor impact on performance, and they were able to keep up with, or even outperform unconstrained ABUs in some of the tested settings. The good performance of normalized ABUs in our tests is in line with [90], who found very similar or identical units[1] to outperform common activation functions in MNIST,

---

[1] With respect to the constraints, Manessi and Rozza [90]'s *affine()* units are equivalent to ABU_{nrm}, and their *convex()* units are equivalent to ABU_{pos}. Unfortunately, the authors did not provide details with respect to their implementation, so we cannot say whether or not the implementations are equivalent.

Table 7.3: Performance (percentage correct) after initialization of scaling / blending weights α at the final values of a preceding run, i.e., after 60 000 steps. Additionally for ABUs: Pre-trained weights normalized at initialization to avoid too low initial scaling. Highest performing treatment of scaling and blending weights per activation function (row) indicated in bold.

Network	SMCN	SMCN	SMCN	SMCN	SMCN
Optimizer	Adam	Adam	Adam	Adam	Adam
Task	CIFAR10	CIFAR10	CIFAR10	CIFAR10	CIFAR10
α init	$\frac{1}{m}$	pre-trained	pre-trained	pre-trained (norm.)	pre-trained (norm.)
α trainable	✓	-	✓	-	✓
αtanh	79.07 ± 0.07	79.60 ± 0.07	**79.71 ± 0.06**	-	-
αReLU	79.23 ± 0.15	79.62 ± 0.11	**79.77 ± 0.09**	-	-
αELU	**82.60 ± 0.06**	79.45 ± 0.15	81.22 ± 0.10	-	-
ABU	**83.12 ± 0.06**	80.93 ± 0.15	82.02 ± 0.15	80.99 ± 0.12	82.88 ± 0.06
ABU$_{nrm}$	**82.82 ± 0.07**	78.88 ± 0.08	81.78 ± 0.14	-	-
ABU$_{abs}$	**83.14 ± 0.08**	79.80 ± 0.24	82.42 ± 0.21	-	-
ABU$_{pos}$	**82.90 ± 0.05**	78.93 ± 0.07	81.45 ± 0.16	-	-
ABU$_{soft}$	82.54 ± 0.07	**82.69 ± 0.05**	81.69 ± 0.08	-	-

CIFAR and ImageNet tasks, using widely used network architectures, such as AlexNet and ResNet-56. In conclusion, while ABUs generally apply adaptive scaling when possible, the ability to learn the function's shape by itself already helps to improve network performance beyond the level of the established activation functions they are comprised of.

7.5 Pre-Training Scaling and Blending weights

Finally, we investigated for both adaptive scaling and ABUs, whether or not the adaptiveness of scaling and blending weights by itself is an important factor for the overall performance of the network, and if the performance could possibly be further improved by using pre-trained weights. To this end, we set up an experiment with two main conditions. In both of them, we initialized the networks with the final scaling or blending weights of a preceding run. We then fixed these values after initialization in one condition, while keeping them adaptive in another. In case of ABUs, it is conceivable that the *shape* of the function at the end of training would be ideal, while the *scaling* may be too low at the beginning of a new run. Therefore, we added two more conditions akin to the main condition and only for unconstrained ABUs, in which we normalized the pre-trained blending weights after initialization, thus keeping the learned shape, while resetting the learned scaling. All tests were based on the default setup (vanilla SMCN, Adam, CIFAR10).

The results are shown in Table 7.3. For αtanh and αReLU, initializing the scaling weights at the predominantly low final values of a preceding run helped to improve performance. In both cases, runs with fixed pre-trained α_i already surpassed the performance of the preceding run, but keeping them adaptive over the course of training led to further improvements. The fact that fixed pre-trained scaling weights yielded an increase in performance suggests that the initial variance of weights in the weight matrices (derived using He initialization), may not have been ideal as initial conditions, despite resulting in pre-activation variances of about 1. αELU, by contrast, substantially lost performance after initialization with pre-trained scaling weights, irrespective of whether or not they were fixed or adaptive throughout the run. With the exception of ABU$_{\text{soft}}$, the same was the case in all versions of ABUs, where fixed blending weights, in particular, led to sizable drops in performance of up to four percent. ABU$_{\text{soft}}$, being the exception to this rule, improved slightly for fixed pre-trained blending weights, but lost performance with adaptive pre-trained blending weights. Normalizing the unrestricted ABU blending weights after initialization with pre-trained values led to improvements over non-normalized pre-trained blending weights, but the default setup with initialization at $\frac{1}{m}$ still performed best. Overall, we found all but one of the tested activation functions to perform best, when the blending weights were kept adaptive, as opposed to fixed after initialization. These results suggest that in both adaptive scaling and ABUs, much of the gained performance is won by keeping the scaling and / or shape adaptive. Moreover, the fact that this applies also to most normalized versions of ABUs indicates that there may not be any single optimal shape for an activation function in a given layer.

7.6 Limitations

In the following, we briefly highlight two noteworthy limitations of this work. Firstly, the ABUs presented here are based on five distinct activation functions, chosen for their prevalence in literature (e.g., ReLU), their standalone performance (e.g., Swish), and to generate a wide range of achievable shapes (e.g., tanh). We believe that a more principled approach to the choice of activation functions used in ABUs might reveal even higher-performing combinations. Due to limited computing resources, we were not able to perform an exhaustive search for optimal combinations of activation functions. Similarly, an in-depth overview of theoretical considerations concerning optimal blends was not provided here, but should be pursued in future work. Secondly, we have so far only evaluated our approach on supervised learning task from the field of computer vision. Future work on this topic should further include experiments based on other applications of deep learning, such as time-series prediction or reinforcement learning.

7.7   Conclusion & Outlook

In summary, we introduced Adaptive Blending Units (ABUs), and analyzed adaptive scaling for common activation functions. We found robust performance advantages of both approaches over established activation functions in a range of tasks and network architectures. In adaptive scaling, the performance advantages could be traced back to stabilized pre-activation statistics during training, mitigating covariate shift. The same behavior was found for unconstrained ABUs, while normalized ABUs reached similar levels of performance without the ability to significantly scale the layer output. Our results suggest that the adaptiveness of the shape over the course of training may play a major role in this, as well, as opposed to simply converging to some ideal shape.

With respect to adaptive scaling, a logical next step would be to introduce a shifting parameter per layer, to allow the network to further optimize the input distributions to the activation functions, and to move this learned normalization in front of the activation: $f(\alpha_i \cdot (x + \beta_i))$. This form of self-organized normalization could be explicitly combined with ABUs, thus detaching the handling of layer statistics from the shape of the activation function. Recurrent networks may be a particularly interesting field of application for self-optimization of layer statistics, as it should, in principle, mitigate some of the issues associated with explicit normalization techniques. Interestingly, adaptive scaling has been discussed for neural populations outside of the field of deep learning and proven helpful in maintaining stable output distributions [120, 82]. Beyond this, an increase in the number of distinct ABUs within a layer may yield further improvements in performance, as well as a systematic search for high-performing sets of activation functions in ABUs.

Chapter 8

Discussion and Conclusion

This concludes the publications included in this book. In the final chapter, we shall give a brief summary of the contributions made to the field of behavioral ethics, embed these into the context of other work that has emerged since the beginning of our work, and highlight open questions in this field. As a guideline for this discussion, we will use the four main questions defined in the introduction chapter.

8.1 Delivering a frame of reference: How do we as humans behave in traffic dilemma situations?

The question how we as humans decide in dilemmatic traffic situations is important as it provides both a frame of reference, and lets us gain insights on where official regulations may be at odds with popular opinion. While we were among the first to publish results on human behavior in traffic dilemmas, a number of publications have since amassed and the picture is completing. In the analysis, we generally distinguish between two kinds of factors with an impact on the outcome of trials. First, features of the situation (sometimes referred to as global preferences) contain all factors that describe the situation as a whole: obstacles, potential victims, expected damages, rule-violations, as well as further contextual variables. Within these, features of the potential victims, such as their age and gender, have received the most attention so far. Second, features of the respondents (e.g., their gender, age, cultural background, or personality traits) may correlate with general behavioral biases and influence the evaluation of particular features of the situation.

In [109] we established that linear models are generally appropriate to model people's moral behavior in dilemmatic traffic situations, and that the valuation of groups of multiple victims is approximately additive, or at least higher than each of the individual valuations. These two findings may seem unsurprising in hindsight, but were important factors to confirm when research on this topic first became relevant. This established the methodological basis for further studies both from our lab and others. The study also found a bias towards a preference for younger people (age bias), and at least hinted at a gender bias favoring females over males. In [112] we focused our trials more on the evaluation of human vs. human victims and used hierarchical models which let us control for various potentially confounding variables. This lead to a clearer detection and quantification of the gender and age biases, showing a moderate advantage for females over males, and a strong advantage for younger over older people. With regards to effects of the "participant features", we found minor differences between males and females with regards to the age bias, as well as some evidence for people with high social desirability to exhibit larger omission and gender biases.

In their large-scale online survey labeled the Moral Machine, [5] analyzed a large number of potentially influencing factors on both the side of the portrayed situation and potential victims (termed global preferences), and individual variations between the respondents [5]. In the analysis of the global preferences, the largest effects were found for favoring humans over pets, larger groups of characters over smaller ones, and younger people over older ones. Further notable effects were found favoring those who behave lawfully, those with a higher social status, the physically fit, females over males, and pedestrians over passengers of the AV. A small effect was also found favoring inaction over action, suggesting that some reluctance to interfere in such a situation is part of our moral intuitions, but is often outweighed by utilitarian considerations. On the side of individual variations among the respondents, small effects were found, for instance, for the respondents' age and gender. These findings generally corroborate the results of our aforementioned studies, as well as others (e.g., [39, 11, 83, 43, 71]). Beyond that, the analysis also looked at cultural differences, distinguishing between western, southern, and eastern cultures. Some notable findings from this analysis include stronger preference to spare pedestrians and the lawful, the absence of any age bias, and little regard for a person's physical fitness in eastern cultures. Southern societies, on the other hand, put a high value on a person's age, their status in society, and are highly protective of females. Finally, western cultures show a strong preference for inaction, potentially a sign of a more deontological understanding of ethics.

One aspect that may require further scrutiny is the effect of lawfulness and / or safety guarantees for pedestrians on the sidewalk. While the Moral Machine found at least moderate effects for sparing lawful pedestrians over jaywalkers, in studies by [74, 39, 11] the protective effect of being positioned on a sidewalk was small, and often outweighed by even minor differences in group size. And while these situations are not identical, they share the distinction between protected unprotected behavior. Differences in the used analyses between these studies make it difficult to compare the effects directly, but there appear to be discrepancies that would warrant a more thorough investigation of our moral judgment when it comes to potentially negligent behavior of pedestrians. Such findings would have practical relevance: As we showed in [113], strict if-then rules lead to unreasonable decisions in cases where (very) low probabilities are met with high stakes. E.g., is swerving onto the sidewalk permissible to save a life, when a minimal but non-zero risk to a pedestrian is involved? In the publication, we concluded that in order to avoid unreasonable decisions, a weighing of expected damages or injuries and the severity of legal transgressions could be performed. Yet how much of a protective factor should be assigned to sidewalks, marked pedestrian crossings, green lights, etc., is an open question.

A factor that is somewhat controversial in this debate is that of preferential treatment for vehicle passengers. Advocates for it argue that self-preservation as an ethical argument is accepted in society [10, 23], and manufacturers have a clear incentive to provide their customers any safety-benefit possible. [13] report that the majority of respondents would prefer AVs sacrificing their passengers for the greater good, yet only a minority indicated they would also be willing to buy a vehicle with this kind of utilitarian programming. The most obvious counter-argument is that a (mandatory) equal treatment of passengers and pedestrians is in societies best interest [48, 86]. On the empirical side, [39] found humans to overwhelmingly decide in favor of the utilitarian option that sacrifices the individual driver. [74] found the perspective from which the situation is presented to have a significant influence on this assessment: From the passenger's point of view, participants chose to sacrifice a group of pedestrians more often than from the pedestrians point of view. Similarly, an external observer's point of view also yielded preferential treatment of pedestrians over vehicle passengers. This finding is echoed by [71], who in a large-scale text-based survey found a pedestrian life to be valued higher than a vehicle driver's life of the same age. Thus, there appears to be a conflict between what is generally perceived as the right moral choice by neutral observers, and what drivers or buyers consider fair. However, focusing on the distinction between an active human driver and a passive passenger of an AV might resolve this conflict: In a driver versus pedestrian scenario, the driver is often assumed as the source of the risk, and may thus carry more of the blame. In an AV-passenger versus pedestrian scenario, however, the AV-passenger may be perceived as entirely passive and thus absolutely innocent with regards to any particular situation. On the other hand, one could also argue that by choosing to use a motorized vehicle, an AV-passenger implicitly consents to a greater mobility risk than a pedestrian does. Disentangling the various factors at play here from an empirical, but also from a philosophical, and finally legal point of view is, to the best of our knowledge, an unsolved task to date.

Altogether, we see a surprising amount of consensus between studies on some of these valuations that appear to constitute the basis for our decision-making in these scenarios. The findings highlight discrepancies between regulations, basic law, and legal necessities on the one hand, and our moral intuitions on the other. They show us where there are large variations between individual members or groups within a society, and they tell us about discrepancies between different cultures and societies. The most obvious example of this is the factor age plays in these decisions: The equal value of all human life, regardless of origin, gender or age is an integral part of many constitutions (e.g., the German Constitution), yet the studies in this field paint a very clear picture that this is not generally reflected in the judgment and behavior of people in western and southern societies. [71] even found an almost equal valuation of individual expected life years in Sweden, which means that the life of a 10 year-old child with an expectation of 70 more life-years to come would be regarded about seven times more valuable than the life of a 70 year-old with a life expectancy of 10 more years. At the same time, [5] found the factor of age to be virtually irrelevant in eastern cultures, highlighting enormous variability between different societies. We thus deem it highly important that these findings be considered in the communication between lawmakers, manufacturers, and the general society, in order to prevent misconceptions, lack of trust in AVs or even public outrage over implemented behaviors. After all, once AVs can be shown to be significantly safer than human drivers, it's in societies best interest to adapt the new technology as quickly as possible, and not be held back by omissions in communication.

8.2 How do we best assess human decisions? What biases are introduced by the assessment?

This question addresses the methodological basis for this kind of research. As pointed out in the introduction, research in other contexts of behavioral ethics has shown significant effects of the presentation or framing of dilemmatic scenarios on the response patterns elicited by respondents. It is thus important to get an understanding of how the measurement might have an impact on the obtained results. In our studies we looked into the level of abstraction, the modality (immersive VR vs. desktop screen presentation), and the length of response time windows. With regards to the level of abstraction and presentation modality, no major effects were found between groups. Minor effects found, for example, with regards to the valuation of different age groups between text-based and naturalistic presentation. These may be linked to differences in the saliency of age discrepancies between potential victims – caused either by a more prototypical "old vs. young" distinction in the cognitive decision-making process, or by a mere failure to visually distinguish between age groups to the same extent in naturalistic displays of the scenarios. Time constraints appear to mostly have an impact on response patterns when they are short, i.e., under four seconds. In [112] we found significant differences between the short (1.x seconds) and medium length (4.x seconds) trials: In short trials, all biases were reduced in magnitude compared to the slower trials. This is consistent with a higher error rate, as one would expect in case of erroneous identification of the potential victims, an interruption of the cognitive decision-making process, or simply erroneous key presses. No significant differences were found between the medium length (4.x seconds) and unlimited trials. The implications of this are mostly relevant for in-motion presentations, which should either be slowed down enough to give participants up to 4 seconds of response time, or paused, even at the cost of immersion.

Beyond these factors, [74] have found effects of the perspective from which the situation in question is shown: When one and the same scenario configuration was shown either from the perspective of a vehicle passenger, a potential victim, or an observing bystander, study participants showed varying judgments of what the right action would be. From the viewpoint of a passenger, participants were more prone to sacrifice the pedestrians than from either of the other two perspectives. In the same vein, [13] found respondents to generally prefer utilitarian be-

haviors over safety-benefits for vehicle passengers, yet they would much
more likely buy a car that provides preferential treatment to its passengers at the expense of pedestrians and other non-passengers. These
findings have implications for the selection of the right perspective for
assessments of potential advantages of vehicle passengers over non-
passengers. It would appear sensible to use an observer's perspective
for these cases, unless otherwise required, and account for the effect
that various perspectives have in the interpretation of results.

[74] further looked into effects of framing, i.e., labelling the vehicle
either as manually operated by a human driver, or automated. No substantial differences were found in the assessment of what would constitute morally right behavior. However, the authors did note a trend
towards allowing AVs more minimization of the overall harm, while
human drivers were expected to uphold individual rights of people to a
higher degree.

Overall we see very comparable results across various presentation
styles, which boosts comparability between different studies and provides researchers with freedom to choose the assessment methodology
without having to account for substantial biases. But what could explain
the differential susceptibility to differing presentation styles between
traffic dilemmas and the more classical thought-experiments from the
literature? Classical vignettes are typically designed to be morally max-
imally ambiguous, and often portray a clash of two moral theories, or
moral intuitions (e.g., minimizing harm vs. non-interference). When
it comes to responding to a traffic-dilemma it appears that value-of-life
based models describe our evaluations quite well (see [109]), and do
not invoke a high level of ambiguity once we have established some
form of internal value system.

8.3 How can we model human behavior in traffic dilemmas, and how do these models relate to decision making in AVs?

We'll start with the simpler first part of this two part question. In a binary forced choice paradigm like the one used in our studies and many
others, regression models akin to those used in [71] stand to reason as
a way to model the outcome of the trials. However, the human brain is
highly complex and we are far from fully understanding the processes
behind decision-making on a neurological basis. First coarse attempts
to understand the underlying processes can be found in psychologically
motivated models like Greene's dual process theory [57] or Cushman's

dual-system framework [28]. But these are very narrowly focused on certain kinds of moral decisions and conflicts between moral theories, highlighting aspects that appear not to be particularly relevant in the traffic dilemma setting. And as such, they do not explain ethical thinking, judgment and decision making as a whole, nor do they really explain the neurological basis for it. The aim of studies like ours, then, is not to postulate or confirm a process model of how our brains make decisions in such situations, but to find statistical correlations and reasonable heuristics that let us predict the outcome, i.e., the final decisions we make. The choices we, therefore, face when constructing a statistical or machine learning model are (1) which variables to observe (and experimentally vary), and (2) what level of complexity and which constraints to equip the model with in order to achieve the best predictive power. Of course, a successful model may give us some insight into how we as humans make decisions, but it should not be understood as a description of neural neural processes.

In [109] we focused on the complexity of the models employed, and established that relatively simple, value-of-life-based regression models are generally appropriate to model our behavior in traffic dilemmas. In [112], we built upon these findings and used hierarchical regression models to find out about and control for various other relevant factors, thus getting a cleaner reading of the magnitude of the main factors. In the statistical analysis of the Moral Machine experiment, [5] used a similar model to ours, as did [71] in a large scale mail-in survey in Sweden. To answer the initial question about the modelling of human behavior, linear value-of-life based regression models are to date a proven and popular method to model moral judgment and behavior in forced-choice traffic dilemmas. They allow us to describe the observations in statistical terms on one side, and to predict a participant's response for new scenarios on the other. Moreover, making such predictions of a participant's expected response under arbitrary combinations of factors is exactly the task that a decision making algorithm would perform in an AV. Thus, provided the factors considered in experimental trials match those the AV can detect in its sensory input, the models derived from such studies could technically be used to make decisions in AVs.

This observation leads us to the second half of the initial question – how do these models relate to decision making in AVs? While technically, such models could be used to make decisions in AVs, we can't ipso facto assume that they should. We shall therefore discuss the following questions: (1) The way the trials are typically formulated in forced-choice studies is considerably simpler than the real world scenarios in which the resulting models would be employed. Does this limit their applicability or utility? (2) Assuming we *could* use value-of-life-based regression models, *should* we? And (3) if we agree to use regression models for decision making, the behavior of the algorithm would be largely determined by the chosen parameters. Should behavioral assessments guide the choice of these parameters?

(1) *Could* regression models be used to make decisions in vastly more complex real-world environments? The trials in the studies discussed are designed to be straight forward and easy to understand. Participants are facing exactly two mutually exclusive options, and it is either stated or implied that (a) the probabilities of a collision are 100% for either of them, and (b) potentially fatal for the victims. As elaborated on in [113], such simple scenarios almost never happen in the real world. Instead, AVs may be facing choices with vastly different and often very low probabilities of colliding with various obstacles, uncertainty about the behavior of other involved parties, and may need to factor in material damages and violations of traffic rules.

But this discrepancy is not necessarily a problem. One could multiply the estimated probability of colliding with an obstacle, the severity of that collision, and the valuation of the obstacle to arrive at the expected loss in value, as suggested by [52]. This could include factors outside of the risk to people as well, such as the risk to animals, a numerical cost value of traffic violations, and projected material damages. As long as the valuations are happening on the same scale, or in the same "currency", any number of factors can be included. It is thus possible to set up a model in which for each behavioral option the expected losses are summed up, and the one with the lowest available loss is chosen [123]. Thus, from a technical standpoint, value-based regression models and algorithms derived from them *can* be used to perform ethical decision making even in real-world scenarios.

(2) *Should* regression models be used for decision making in AVs? The previous section outlines the technical feasibility of employing regression models in the real world. The more challenging question is, however, not whether we *could*, but whether we *should* use regression models for ethical decision making in AVs. A very strong point in favor of such models is their ability to deal with uncertainties and a probabilistic conception of the world and predicted future events. As discussed in detail in [113], we deem this a necessary prerequisite for ethical decision making systems. Moreover, they are compatible with a number of demands stated in the same publication, e.g., reasonableness, robustness, and transparency of ethical evaluations, and sensible violations of traffic regulations when the situation calls for it.

And while such a model would do justice to the complexity of reality, it would also shift some of the burden to the sensory systems to provide risk assessments with the highest precision possible. However, a precise assessment of risk plays an important role in making good decisions in driving regardless of the type of decision making algorithm tasked with processing the assessments further. Once enough cars are equipped with advanced sensory hardware, their collective experience may give rise to fairly precise risk assessments in many common and less common situations. An initial inability to make precise assessments of risk, on the other hand, should not be used as an excuse to use overly simplistic decision making algorithms at the cost of safety and fairness in traffic.

An interesting counter-argument to utilitarian decision making systems is provided by [23], who argues that "[w]ithin the framework of a liberal legal system that recognises humans as free agents who have rights and duties, maximising the function of social utility does not justify harmful interference into a person's legal sphere. There is no holistic entity whose interests have to be maximised, even in a situation of necessity." The author performs an in-depth analysis of five dilemma situations, providing precise assessments of legal justification from a deontological perspective. And while his analysis appears perfectly valid for the framework of criminal law to date, it inadvertently displays this framework struggles to indicate practical solutions to ethical decision making in AVs. The legal analysis offered relies on a discrete action space, i.e., a finite number of distinctly different options, and, more importantly, it requires the outcome of the decision to be known before the fact. As we laid out in [113], at least the latter requirement is impossible to meet in the real world. Thus, while the author recognizes that "it is not enough to find a plausible way of laying foundations for the impunity of the person acting in a tragic incident, but a clearly defined rule must be established that aims to be generally valid for the dilemmas that may arise" [23], his suggestion does not provide any practical guidance with regards to the programming of AVs. The analysis further indicates that minor changes in the assessment of the situation, or in the predictions it makes, could change which of the behavioral options would be legally favorable, doing away with any notion of robustness. Moreover it would require each of thousands of possible situations to be evaluated individually, which is most likely infeasible. Adding to this that the author concedes some of his conclusions are, to a degree, subject to individual judgment and could be contested by other scholars, inconsistencies would be likely, and the requirement of transparency of the decisions would not be met. It may thus be the case that the current framework of criminal law may not be capable of delivering a consistent and complete set of instructional rules suitable to guide the programming of AVs.

On the other hand, approaches like risk management [52] and utility functions [123] *would* provide a solution that lends itself to implementation in practice. Such systems don't weigh options in a symbolic manner, or from a deontological standpoint like legal analyses under the framework of criminal law, but it is conceivable that the model and its parameters could be set up such that the system would come to the "correct" conclusions in a number of defined test cases, while still being able to work based on probabilistic assessments of the situation. In a sense, it may be able to satisfy the law, while ensuring reasonableness, robustness, and transparency at the same time. We thus believe that despite their utilitarian roots, regression model-based decision making algorithms constitute a viable option to solve ethical decision making in AVs and need to be considered for this purpose by lawmakers.

(3) Should behavioral assessments guide the choice of parameters in real world decision making algorithms? This question loops back to the behavioral experiments conducted by us and others, and addresses the value of these insights for decision making algorithms in practice. The short answer is that most of the insights into our intuitive moral behavior in traffic dilemmas are relevant to the communication between lawmakers, manufacturers and the public, but should not be represented in the ethical programming of a car in any way. This view is in line with sentiments in the medical community regarding empirically assessed criteria for organ recipient rankings [15].

Legal complications aside, using the value-determining parameters estimated in behavioral studies for utility or loss assessments in decision making algorithms would in many cases constitute committing an *is/ought fallacy*. This fallacy describes the assumption that because things are a certain way, they should always be this way. In the case of moral behavior it is particularly enticing to do so: Since there is no absolute right or wrong, what we perceive to be ethically acceptable or not is largely dependent on our own intuitions and deliberations, which in turn are strongly influenced by culture and personal experiences. One might consider that moral doctrines and theories may in many cases be conceived after the fact as a justification for one's own moral intuitions, rather than giving rise to these intuitions in the first place. It is also somewhat controversial whether or not a person, such as a moral philosopher, can be a true moral expert in the sense that their moral judgments and evaluations should be considered superior to those of non-scholars. Based on these notions, one might find that if a majority of people intuitively arrive at a similar moral judgment, then this judgment represents what ought to be. However, such an assessment would be highly problematic. Judgments assessed in specific contexts tend to be context- or even framing-specific, and therefore inconsistent over varying contexts, and difficult to justify. As a consequence, they would be rightfully superceded by more general moral values and norms that are integral to society, when these are at odds with each other. Case in point, the fact that a large majority of people in western societies intuitively grant younger people higher protection, or higher value, does not in itself justify the claim that this is the morally right thing to do for everyone, and in particular a machine. On the legal stage, it would certainly not justify bypassing constitutional laws which grant all people the same rights, regardless of their age, sex, ethnicity, religion or any other personal features.

Therefore, intuitive biases that discriminate based on such personal features are essentially precluded from being considered in the decision logic of AVs. Other behavioral biases that can be assessed in traffic dilemma studies, however, may have practical relevance: In its current state, the law either allows or prohibits certain actions. Special circumstances may make infringements quasi-permissible, but such cases are typically decided in court after the fact, rather than specified a-priori. Weighing up the utility, loss, or justifiableness of various options in a dilemma situation, however, requires precisely such a-priori specifications of what would constitute a justified violation of the law. In a regression model-based decision making system, finite parameter values would have to be set for a number of factors that can become relevant in a critical situation. For example, defining sidewalks or marked pedestrian crossing as protective zones requires a quantification of the extent of this protection. Moreover, utility costs for red light violations, driving into the opposite lane, and other violations of traffic regulations would have to be defined, to specify the amount of damage prevented, or risk avoided that would justify them. Since it may prove difficult to base such quantifications on existing law or legal practice, behavioral experiments could deliver a frame of reference for the order of magnitude of such parameter values.

Overall, we conclude that algorithmic ethical decision making in AVs *could*, and possibly *should* be based on regression models. Where parameter values have to be set for which existing laws do not provide any guidance, observations from behavioral studies could deliver a frame of reference.

8.4 What challenges remain on the road to automated ethical decision making in AVs?

While a lot of progress has been achieved in recent years both on the technical, as well as the ethical side of vehicular automation, some key challenges remain to date.

As evidenced by the preceding discussion, the question whether a decision making system that is rooted in utilitarianism can be legally acceptable under certain conditions constitutes possibly the largest hurdle on the way to automated ethical decision making. Different jurisdictions may arrive at different conclusions in this question. Those granting implicit or explicit permission to such systems may tie this permission to a number of rules or constraints for the decision making logic. Implicit permission could be realized by defining a certification standard that includes a number of test scenarios to which the system has to respond in a certain way. This way, the regulation could make sure that the system's decisions are in line with those derived from deontological reasoning in defined cases, while allowing for a notion of reasonableness, robustness, traceability and transparency in situations where the outcome is less clearly defined at decision time. Under explicit permission, one would further have to consider whether or not parameters or parameter ranges should be defined by the regulation itself. Jurisdictions rejecting utilitarian-rooted decision making, on the other hand might be forced to develop alternative solutions that still take practical constraints into account, such as uncertainties in situational assessments and predictions.

Other open challenges in this realm concern the previously discussed quantification of behavioral biases, e.g., whether or not these should be provided by law, and whether personalized ethics settings (PES) could be an option to avoid the potential problem of algorithm aversion [32]. We discussed these issues in detail in [113]. Finally, in practice, automated ethical decision making also requires accurate assessments of the immediate environment of the car, including the recognition of involved entities and a prediction of their movement or behavior in the immediate future. While not being strictly a part of the decision making itself, an accurate assessment of the situation provides the basis for any adequate decision making algorithm. The primary technology used to provide the required information processing in AVs is machine learning. In particular, deep learning based processing of camera images and fusion of image data with other sensors are the central elements to successful recognition of relevant elements in the environment. Progress in this field is rapid, and each update to a vehicles' ADAS or automated driving software modules pushes the performance boundaries. Nonetheless, getting camera-based scene recognition to a level that at least equals the performance of human drivers is likely the biggest technological challenge on the way to automated driving on level 4 and 5. We make a minor contribution to this development with our publication on adaptive activation functions [111], which could lead to improved assessments of critical situations and, therefore, better ethical decisions in the future.

8.5   Conclusion

To conclude, since ethical questions regarding AVs first gained widespread attention around 2015, the research community has extensively explored and analyzed the most challenging questions, engaged in fruitful scientific discourse, and delivered a solid philosophical and scientific basis for necessary ethical and legal decisions by lawmakers and manufacturers.

As part of this research community I believe that that the work of my colleagues and I has a part in this achievement. This book represents a multi-facetted approach to analyse and address core challenges related to automated ethical decision making in AVs. Over the course of five publications and a subsequent discussion, we have analyzed and discussed philosophical and legal aspects of the issues, employed psychological or behavioral experiments in virtual reality, have applied and evaluated machine learning models as a potential solution to the design of decision making algorithms, and contributed to the underlying sensory processing stage of AVs with methodological research in the field of deep learning. Altogether, I believe that this book provides a structured and comprehensive overview of the state of the art in ethical decision making for AVs, includes important implications for the design of decision making algorithms in practice, and concisely outlines the central remaining challenges on the road to a safe, fair and successful introduction of automated vehicles into the market.